Air Pollution Abatement and Regional Economic Development

Air Pollution Abatement and Regional Economic Development

An Input-Output Analysis

William H. Miernyk
John T. Sears
West Virginia University

Lexington Books
D.C. Heath and Company
Lexington, Massachusetts
Toronto London

Library of Congress Cataloging in Publication Data

Miernyk, William H.
Air pollution abatement and regional economic development.

Bibliography: p.
1. Air–Pollution–Economic aspects–United States. 2. Regional planning–United States. 3. Interindustry economics. I. Sears, John Thomas, 1938- joint author. II. Title.

HC110.A4M53 301.31'0973 74-8234
ISBN 0-669-94656-7

Published simultaneously in Canada.

Printed in the United States of America.

International Standard Book Number: 0-669-94656-7

Library of Congress Catalog Card Number: 74-8234

Contents

List of Figures

List of Tables

Preface

Widespread concern about environmental degradation is a fairly recent phenomenon. A decade ago there was little in the economic literature that dealt with the problem of air or water pollution. At that time the emphasis was almost entirely on growth and development. Today, there is voluminous literature on the economics and technology of environmental management. The selected bibliographies at the end of this book cover only a fraction of what has been written in recent years. This being the case, what is the justification of another study of air pollution? How is this study differentiated from others that have dealt with the same general issue?

First, it is not simply an economic study; it is an amalgam of economic and engineering analysis. It is an interdisciplinary study in the precise meaning of that term.

Second, the focus is on regional economic development, not on environmental management *per se*. The objective was to determine what effects, if any, clean air legislation would have on the programs designed to stimulate the development of regions characterized by economic stagnation or at best slow rates of economic growth.

The analytical basis of the study is a dynamic, regional input-output model. Input-output models have been used in earlier studies—particularly at the national level—to analyze the economics of pollution and its abatement. Most of them have been static models, however, with the analysis limited to the effects of pollution abatement on operating costs. Since this study is based on a dynamic model it deals with the capital requirements of pollution abatement. The impact of meeting clean air standards on operating costs are discussed fully in this report. But we have also provided detailed estimates of the new and replacement investments that will be required to meet air quality standards established by the Clean Air Act of 1970.

Finally, one chapter of this book deals with the regional economic impact of coal gasification. We first became interested in coal gasification as a partial solution to the problem of meeting clean air standards. At the time we started to look into the economic feasibility of converting coal to gas, there was little concern about the issue of energy supply. This part of the study may be particularly timely, however, in view of the international "energy crisis" precipitated by the outbreak of hostilities in the Middle East in October 1973.

The empirical portions of the study are limited to a single state—West Virginia. And it deals only with the cost of abatement of air pollution from stationary sources. We hope, however, that the findings will be more generally applicable and in particular that the method used to measure the cost of pollution abatement will be helpful to other scholars dealing with the same or similar problems.

From a methodological point of view this is not a complex study. What it involves, essentially, is the reconstruction of 17 of the 48 sectors in the West Virginia input-output model. This meant the collection of new data, however, and data collection invariably proves to be a time-consuming activity. In an effort to keep the report readable we have presented only the summary results of coefficient changes in the text. Complete details on a sectoral basis are given in the appendix tables.

This study was conducted over a period of more than two years. The first phase was supported entirely by the Regional Research Institute of West Virginia University, but during the last two years the project received substantial support from the Economic Development Administration of the U.S. Department of Commerce. Completion of the study depended upon the cooperation of public utilities and a substantial number of industrial firms in West Virginia. Although they must remain anonymous, we wish to thank the representatives of firms that provided us with data. In particular, we thank the firms that permitted staff members to visit their establishments to inspect abatement equipment and to discuss the costs and problems of retrofitting on the spot. We are also grateful to members of the staff of the West Virginia Air Pollution Control Commission for providing us with data and related information.

Special thanks are due to Professor Arthur M. Squires, of the City College of the City University of New York, to Paul H. Gerhardt, assistant to the Director of the Implementation Research Division of the Environmental Protection Agency, and to Professors Richard T. Newcomb and Duane R. Skidmore, of West Virginia University, for their helpful comments on an early draft of our chapter on the simulation of low-BTU coal gasification.

It is a pleasure to acknowledge the important contribution to this study made by the following graduate assistants: Luther C. Thompson, Frank A. Giarratani, Wayne H. Duff, Robert T. Benton, and W.Y. Soung. They, in turn, were assisted by Melissa Wolford and Lorena Goodson. Those familiar with the mode of operating academic research programs will know that the graduate assistants are responsible for much of the difficult work involved. We are pleased, however, that some of them were able to further their professional development while working on this study. Wayne Duff completed his work for a Master of Science degree in Chemical Engineering while a member of the project staff. Frank Giarratani and Luther Thompson have gone on to do independent work on the price effects of air pollution abatement using the model and data of this study.

Last, but far from least, we want to thank Linda Yazvac who was responsible for preparation of the manuscript, including most of the original sketches and figures. In typing the manuscript she was ably assisted by Jean Gallaher. The final art work was ably handled by Peter L. Heiss, artist with the WVU Office of Publications.

It is evident that this has been a team project. Without the assistance of those named above it would not have been completed. We must add the caveat, however, that the authors alone bear responsibility for the final results.

Air Pollution Abatement and Regional Economic Development

1 Introduction

Regional Development and Environmental Protection Legislation

On May 1, 1961, President John F. Kennedy signed the Area Redevelopment Bill, the first major legislative act of his administration. The new act had a long and involved legislative history that will not be discussed here.[1] But this legislation marked a new departure in American public policy. For the first time the federal government pledged itself to take an active part in efforts to revitalize depressed industrial areas. The legislators who drafted the act also hoped it would do something to improve economic conditions in lagging rural areas. This hope was based in part on the experience of the Tennessee Valley Authority established in the 1930s. But the new law was nationwide in scope, and its objectives differed somewhat from those of the TVA.

The Area Redevelopment Administration, a separate agency established by the act, had a brief and rather stormy history. Under the Public Works and Economic Development Act of 1965, it was replaced by the Economic Development Administration, which is part of the Department of Commerce.[2] That year also marked establishment of the Appalachian Regional Commission under the Appalachian Regional Development Act of 1965.[3] By the mid-1960s, the United States had made major commitments to programs of regional economic development.

The decade of the sixties was one of sustained economic growth for the United States. But for a variety of reasons discussed in the books by Levitan, Hansen, and Newman cited above, the residents of a number of regions in the United States failed to share in the benefits of national economic expansion. Their representatives in Washington were joined by other members of Congress who were concerned about the failure of the market system to reduce geographic inequities in income and economic opportunity. This coalition was able to marshall support for the regional economic development programs mentioned above.

Historically, economic development has been accompanied by undesirable side effects, particularly air, water, and solid-waste pollution, and congestion. But while such "negative externalities" have long been recognized in the literature of economics, relatively little has been done about environmental problems until recently. There are some earlier antecedents, but the first federal efforts to control air pollution are found in the Air Quality Act of 1967.[4] An important feature of this act was its regional approach to the establishment and enforcement of air quality standards. It provided for the designation of Air Quality Control Regions (AQCRs) and the establishment of criteria by which states were to develop their own standards of air quality. States were expected to establish implementation plans and to set specific emission levels for individual sources. They also were to provide time tables for compliance. The federal government retained authority to set standards if a state failed to do so.[5] The next step was the Clean Air Act of 1970, which directed the administrator of the Environmental Protection Agency to set ambient air standards for the nation.

Within the span of a decade, the federal government has attempted to implement two policies that may appear to have conflicting objectives. On the one hand, it has undertaken to stimulate *new* economic development in several regions charcterized by slow growth or economic stagnation. At the same time, the federal government has established national air quality standards. It has required the states to take whatever steps are necessary to reduce air pollution that equals or exceeds specified ambient air quality standards. The objective of this policy is to reduce to acceptable levels one of the negative externalities or undesirable side effects of economic development.

Are the two policies described in the preceding paragraph actually contradictory? Is the federal government using public funds to stimulate economic development in some regions while at the same time enacting legislation that might hinder further industrial development in those regions? It is at least a testable hypothesis that legislation designed to improve air quality will have an adverse effect on economic activity in some regions.

What would happen to the West Virginia economy, for example, if stringent air quality standards resulted in cutbacks in production and employment in one or more industries generally classified as heavy polluters? The direct effects would be immediately apparent. But the indirect effects might be more difficult to observe. These effects can be estimated, however, through input-output analysis if an input-output model exists for the region in question. Such a model has been constructed for West Virginia. The economic effects of compliance with clean air standards reported in the chapters that follow are based entirely on this model.[6]

The details of the West Virginia model and the way in which it has been used to estimate the cost of clean air in West Virginia are given in Chapter 2. For present purposes it will suffice to point out that the model describes the West Virginia economy in considerable industrial detail; it records interindustry transactions as well as sales to final consumers both in-state and elsewhere. This permits us to estimate the indirect as well as the direct effects of any change in sales to final consumers on all sectors of the regional economy. In addition, there is a dynamic version of the model that includes tables of expansion and replacement capital coefficients. This version allows us to estimate the effects on expansion and replacement investment of changes in sales to final consumers. Finally, the model includes a set of tables that shows the distribution of employment by occupation and industry, together with tables of labor coefficients. These tables may be used to estimate the employment impacts of changes in sales to final consumers. If a dynamic model is stable (in a mathematical sense), it can be used to estimate the effects on output, investment, and employment of *negative* changes in sales to final consumers as well as positive changes. In brief, it can be used to estimate the direct and indirect effects of economic decline as well as the direct and indirect effects of economic growth.

During an early stage of this study we were interested in measuring the total effects on output, employment, and investment of relatively modest annual declines in production in two of West Virginia's largest industries–steel and chemicals–both of which are considered to be fairly heavy polluters. We did not suggest that compliance with the 1970 clean air standards would mean that such declines were inevitable. We thought it would be interesting, however, to measure the impact on the state economy of an annual decline of 1 percent over a ten-year period in the final sales of primary metal and chemical products manufactured in West Virginia. We assumed that this reduction would be the result of efforts to meet clean air standards. This exercise was intended to show nothing more than the cumulative effects on two specific industries–and the regional economy as a whole–of relatively small annual declines in production.

Because of the high degree of interdependence between the selected industries and other sectors of the West Virginia economy, even the relatively modest declines in annual sales that were specified would have a greatly magnified effect on those specific industries. In the case of chemicals, for example, an annual decline in sales to final consumers of 1 percent would result in a reduction in the total output of the West Virginia chemical industry of more than 38 percent over a ten-year period. The corresponding decline in primary metals would be 32 percent. Because of the importance of these two sectors to the West Virginia economy, the state's gross product would drop by 11 percent as a result of the hypothesized contraction of these two basic industries.

The employment impacts on the state's economy would be slightly less, but in the event of such a decline, total employment in West Virginia would drop by almost 7 percent. Finally, because both of these industries are quite capital-intensive the impact on investment would be large. In the two declining industries there would, of course, be *disinvestment.* Other sectors, which sell to the declining sectors, would also reduce their expansion capital spending. If, in fact, the sales of West Virginia's primary metals and chemical industries declined at an annual rate of 1 percent, there would be a cumulative reduction in investment in West Virginia of almost 30 percent. Internal investment would drop even more, or by 35 percent, while out-of-state investment would decline by slightly less than 20 percent.

Even relatively small negative changes in final sales by two of West Virginia's basic industries would have a significantly large adverse impact on the state economy. This impact would vary from sector to sector, but the total impact on the state would be severe.[7] In a state not yet fully recovered from a precipitous earlier decline in one of its major economic activities–coal mining–concern about the potential adverse effects of pollution regulations is understandable. Fortunately, the empirical findings reported later in this study show that the economic decline discussed briefly above is purely hypothetical. We found no evidence that the industries basic to the West Virginia economy are planning to curtail output in order to comply with clean air regulations. This finding does not rule out the possibility that investment in abatement equipment could slow down growth *rates*, although again we found no evidence that this might actually be the case.

National and Regional Air Quality Standards

Before 1967, federal law allowed individual states and local governments to set air quality objectives. Section 109 of the Clean Air Act of 1970 provided for the Environmental Protection Agency to establish federal air quality standards. This law also provides national emission standards for major new pollution sources [Section 111(a)]. A new source is one where construction or modification began after the federal emission standards had been established. States retain the responsibility for developing implementation plans. As noted earlier, however, the federal government reserves the authority to establish its own plan if a state fails to exercise its responsibility [Section 110(c)].

National primary and secondary air quality standards were established in the early seventies on the basis of criteria studies provided for by earlier legislation. These air quality criteria reflected the latest empirical work available for determining the effects that pollutants may have on human health and welfare. Standards have been established for six pollutants, sulfur oxides, particulate matter, hydrocarbons, carbon monoxide, nitrogen oxides, and photochemical oxidants. The standards for sulfur oxides and particulate matter–the only two pollutants considered in this study–are given in Table 1-1. The other pollutants are not considered here because no state implementation control plan was established for them.

Two classes of national ambient air standards have been established: primary and secondary. The primary standards are set at levels sufficient to protect human health with an adequate margin for safety. Secondary standards are more stringent and are intended to protect the public welfare–that is, protect vegetation, crops, and property. These standards are established on the basis of empirical work cited in the criteria documents for the pollutants in question (see Table 1-2).

Table 1-1
Federal Air Quality Standards

	Primary	Secondary
Particulates		
In micrograms/cubic meter		
Annual geometric mean	75	60
Maximum 24-hr. concentration*	260	150
Sulfur Oxides		
In micrograms/cubic meter		
Annual geometric mean	80	60
Maximum 24-hr. concentration*	365	260
Maximum 3-hr. concentration*	–	1300

*Not to be exceeded more than once a year.

Source: *EPA Regulations on National Primary and Secondary Ambient Air Quality Standards.*

In addition to the standards shown in Table 1-2, the Clean Air Act of 1970 classified several substances (asbestos, beryllion, and mercury), not considered in this study, as hazardous. The Environmental Protection Agency has authority under Section 112 of the act to set emission standards for these substances. It is also the policy of the EPA to prevent degradation of air quality in areas where the air is already cleaner than that called for by national standards.

The West Virginia Implementation Plan

Under the 1970 act, the individual states must develop implementation plans by which national standards are to be met. Each state is to establish specific standards for given sources or classes of sources so that the Air Quality Control Regions (AQCRs) meet the national standards. The factors considered in establishing these regions are meteorological and topographical. Urban-industrial and jurisdictional boundaries are also considered.

West Virginia has been divided into 10 regions for the purpose of the West Virginia Plan. Of these regions, 4 are parts of interstate control regions and include portions of Maryland, Kentucky, and Ohio. Each region is classified by priority for each of the pollutants for which standards must be met. There are three priority classifications: Priority I AQCRs are those which do not presently meet the primary standards for the pollutant in question; Priority II regions meet the primary, but not the secondary standards for a given pollutant; and Priority III regions are those that already meet the secondary standards. Area reduction strategies for specific pollutants are not required for Priority III

Table 1-2
Criteria for Determination of Federal Air Quality Standards

	Lowest Level for Which Effects Have Been Noted on:	
	Human Health	Welfare*
Particulates		
In micrograms/cubic meter		
(Annual geometric mean)	80	60
Sulfur Oxides		
In micrograms/cubic meter		
(Annual geometric mean)	115	85

*Effects on materials and vegetation.

Source: *Air Quality Criteria for Sulfur Oxides* and *Air Quality Criteria for Particulate Matter*, U.S. Department of Health, Education and Welfare.

regions, but reduction plans for emission from specific plants may be required in order to maintain present air quality.

All AQCRs in the state are classified Priority II with respect to all air pollutants for which national standards have been set except particulate matter and sulfur oxides. The priority classification for particulates is given in Table 1-3 and for sulfur oxides in Table 1-4. The regional boundaries are shown in Figure 1-1.

The West Virginia Implementation Plan designates one region for each pollutant as an "example region." These are Region IV, Kanawha Valley Intrastate AQCR, for particulate matter; and Region I, Steubenville-Weirton-Wheeling Interstate AQCR, for sulfur oxides. Control strategies developed for these regions are applied to other regions with similar priorities for the pollutant in question.

Generally, the procedure followed for developing the control strategy was to compile an emission inventory for a particular pollutant, then estimate the total emission reduction necessary to meet the national ambient air standards. Specific emission standards for a given source were developed on the basis of simulation models that provided a method for estimating the spatial distribution of the pollutant for a given source or class of sources.[8]

The concentration of emissions is a function of the emission rate, height of emission, and topography, as well as wind speed and direction. The models used these data to estimate diffusion of the pollutant, and this information provides the foundation for present emission standards. Table 1-5 gives the schedules for compliance with the emission standards as developed in the West Virginia Implementation Plan.

Table 1-3
Priority Classification of AQCRs for Particulate Matter

Priority Classification	Federal Air Quality Control Region for West Virginia	
I	Region I*	Steubenville Weirton Wheeling
	Region II*	Parkersburg Marietta
	Region III*	Huntington Ashland Portsmouth Ironton
	Region IV	Kanawha Valley
	Region VI	North Central W. Va.
	Region VII*	Cumberland Keyser
II	None	
III	Region V	Southern W. Va.
	Region VIII	Central W. Va.
	Region IX	Allegheny
	Region X	Eastern Panhandle

*Interstate Regions

Source: Section 2.08, Series X, *West Virginia Administrative Regulations*, APCC (1972).

Table 1-4
Priority Classification of AQCRs for Sulfur Oxides

Priority Classification	Federal Air Quality Control Region for West Virginia	
I	Region I	Steubenville Weirton Wheeling
	Region VII	Cumberland Keyser
II	Region II	Parkersburg Marietta
III	All other regions	

Source: Section 2.08, Series X, *West Virginia Administrative Regulations*, APCC (1972).

REGION	PRIORITY PARTICULATES	PRIORITY SULFUR DIOXIDE
I	1	1
II	1	2
III	1	3
IV	1	3
V	3	3
VI	1	3
VII	1	1
VIII	3	3
IX	3	3
X	3	3

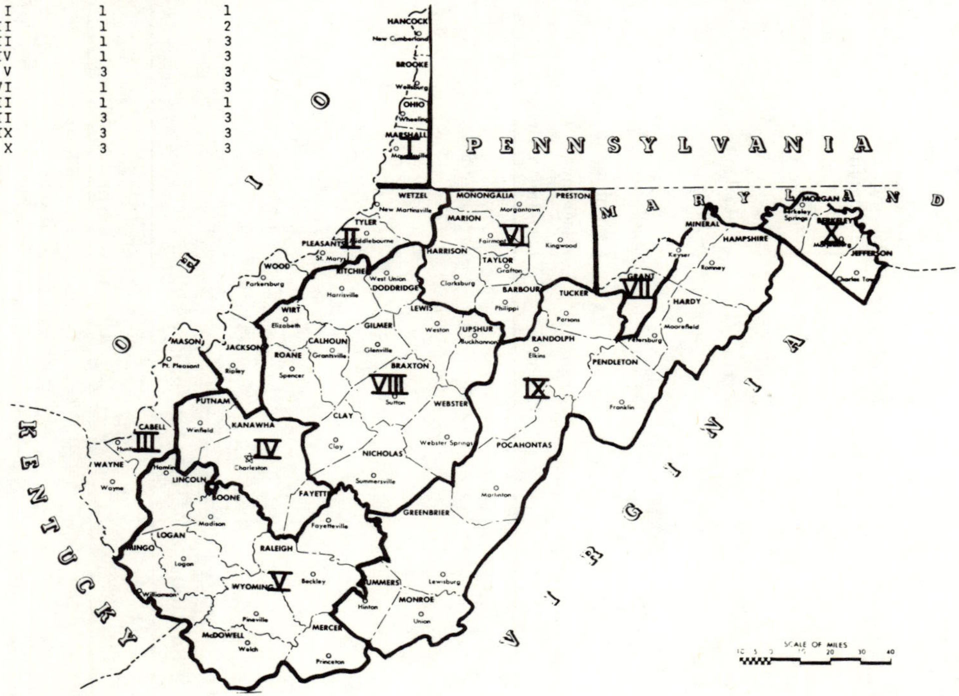

Figure 1-1. West Virginia Air Quality Control Regions

Table 1-5
Compliance Schedules for Emission Standards

	Primary Standard	Secondary Standard
Control of Particulate Matter	June 1975	June 1977
Control of Sulfur Oxides*	June 1975	June 1978

*Holds for all regions except Kanawha Valley (Region IV), for which emission standards became effective January 1, 1973.

Source: *West Virginia Implementation Plan*, Section 6.06.

The next four chapters deal with the methods of estimation, and estimates of the cost of compliance with the standards described above. There are two major air pollutants in West Virginia, particulates and sulfur oxides. One chapter is devoted to each, and the costs of abatement when the two are combined are given in Chapter 5.

We did not limit this study to potential negative impacts of clean air legislation. We also investigated the possibility that management of the environment could have a positive effect on the regional economy. Chapter 6, which discusses the economic aspects of coal gasification, deals entirely with this possibility.

A word should be said about the limitations of this study. It deals only with implementation of the Clean Air Act of 1970, and only with the abatement of industrial emissions from stationary sources.[9] It does not go into water or solid-waste pollution, and it does not touch on the problem of air pollution caused by automobiles or other moving sources. But the method used to make the study could be applied to other types of industrial pollution problems.

Finally, while this study deals only with the economy of West Virginia, we feel that many of the findings are not limited to this state; they should have much broader applicability. The detailed data should be applied to other regions only with great care because each region has a unique industrial structure, and the input-output coefficients for West Virginia apply only to this state. When differences in industrial structure are taken into account—particularly those between in-state and imported inputs—it should be possible to make adjustments in our sectoral cost estimates so that they could be used as approximations of the cost of industrial pollution abatement in other states. Also, the *general* findings about the relationship between regional economic development and environmental management should be as relevant to other regions as they are to this state. It is in this sense that we feel our findings are broadly applicable.

2 Method of Analysis

The West Virginia input-output model was used extensively to answer the questions raised in Chapter 1. Only a brief sketch of the model will be given in this chapter, since it has been described in complete detail elsewhere.[1] The particular use of an input-output model to analyze the economic problems of pollution described in this chapter is, as far as we know, unique. But this is not the first time that input-output models have been used for the economic analysis of pollution problems.

As early as 1965, Cumberland had suggested that an input-output model could be used to measure both environmental benefits and environmental costs.[2] Even before the publication of Cumberland's paper, however, there had been two large-scale investigations of the economic consequences of water pollution using input-output analysis. One of the water studies covered the state of California; the other covered the Colorado River Basin, which includes parts of six western states.[3]

Extensive work on pollution using input-output methods has been done by Wassily Leontief, the inventor of input-output analysis.[4] And long-range estimates of the cost of abatement of various types of pollutants at the national level have been made by Ayres, Gutmanis, and Shapanka.[5] Kohn has published a number of papers on the economics of air pollution control that are listed in the annotated bibliography at the end of this study. He has used input-output analysis in conjunction with linear programming to develop a model designed to minimize the incremental cost of pollution control.[6] One objective of the present study is the measurement of incremental costs of pollution abatement, both direct and indirect. No effort has been made, however, to develop an economic model that would minimize these costs.

A Concise Description of the Input-Output Model

A regional input-output model, such as the one constructed for West Virginia, is a small-scale version of a national input-output model. A static, open model consists of three tables. The first, which is the basis of all input-output analysis, is called the *transactions* table. This is a square matrix listing all of the sectors that make up the economy along the left-hand side while an identical listing is provided across the top. The main body of the transactions table describes all of the productive activities in the economy in terms of two-way flows. Reading across a row one sees, for example, the dollar volume of sales of coal to steel. Reading down a column one sees the purchase of coal by steel, and so on for every sector of the economy. A number of columns are added to the right-hand side of this matrix representing sales to final consumers. This set of columns, referred to collectively as *final demand*, will include at a minimum columns showing sales to households and government agencies, as well as export sales. Typically, there is also a single column summarizing total sales by each sector to final demand.

At the bottom of the square matrix there is a set of rows indicating *primary inputs*. These represent purchases from households, imports, and payments to government. There also may be a row showing depreciation offsets to gross capital formation if there is an investment column in final demand. Again, there generally will be a single row summarizing all primary payments under the heading of *value added* by each sector.

The largest part of the transactions table consists of the *processing sectors*. In the case of national tables, these may run to several hundred. The West Virginia transactions table consists of 48 individual sectors. While this is not a large number, it represents the largest amount of disaggregation that was feasible at the regional level where disclosure rules and other statistical constraints prohibit further disaggregation.[7]

The second table in the static, open model is called the table of *direct input coefficients* or in the case of national models, *technical coefficients.*[8] The input coefficients are calculated by dividing the processing sector entries of each column by the total gross outlays of that column (i.e., the sum of both the processing and final payments portions of the table). The resulting set of coefficients may be thought of simply as the inputs, expressed in cents, required from each of the processing sectors per dollar of output of the sector represented by a given column.

The third table is derived from the second table of input coefficients. It is referred to, after the inventor of input-output analysis, as the *Leontief inverse.* This table is calculated by subtracting the matrix of input coefficients from an identity matrix that, in effect, increases the output of each sector by one dollar, and inverting the resulting difference matrix. The typical Leontief matrix, which does not include households as one of the processing sectors, will have entries greater than one along the main diagonal and smaller than one elsewhere in the table. This matrix is the general solution to the system of equations that make up the static, open model. It permits the model to be used to make forecasts and impact studies.[9]

The West Virginia model is somewhat more complex than the one described above since it is a dynamic model. In addition to the matrix of input coefficients (typically called the A matrix) it includes a matrix of *expansion capital coefficients.* This matrix shows the expansion investment required per unit increase in a given sector's *capacity*. It also includes a second matrix of *replacement capital coefficients* derived from the matrix of expansion capital coefficients. Finally, it includes a matrix of labor coefficients that can be used to estimate the employment impacts of changes in output projected by the basic input-output model.

In general, an input-output model, whether dynamic or static, is designed to answer the question: What will be the effect on each sector of the economy of given changes in final demand? Because the West Virginia model includes capital and labor coefficients it can supply additional information dealing with the investment impacts and employment impacts of specified changes in sales to final demand. The West Virginia model can be described compactly in terms of the following notation:[10]

$$\mathbf{X_t} = \mathbf{A_t X_t} - \mathbf{DX_t} - \mathbf{B}(\mathbf{X_t} - \mathbf{X_{t-1}}) = \mathbf{Y_t}, \tag{2.1}$$

where:

$\mathbf{X_t}$ = a vector of output in year t,
$\mathbf{A_t}$ = the matrix of input coefficients for each year,
$\mathbf{D}$ = the matrix of replacement capital coefficients whose elements are assumed to be constant,
$\mathbf{B}$ = the matrix of expansion capital coefficients with constant elements, and
$\mathbf{Y_t}$ = the final demand vector in year t.

The input and capital coefficients in the West Virginia model were calculated from survey data collected for the base year (1965). The **A** matrix was projected to 1975 using the "best practice" technique.[11] As indicated above, both the **B** and **D** matrices are assumed to be stable over the projection period. The other basic data requirement for this model is a vector of initial output levels for each industrial sector in the year t–1. An identity matrix is added, and (2.1) is rearranged to obtain:

$$(\mathbf{I} - \mathbf{A_t} - \mathbf{D} - \mathbf{B})\mathbf{X_t} + \mathbf{BX_{t-1}} = \mathbf{Y_t}. \tag{2.2}$$

Solving for annual output yields:

$$\mathbf{X_t} = (\mathbf{I} - \mathbf{A_t} - \mathbf{D} - \mathbf{B})^{-1}(\mathbf{Y_t} - \mathbf{BX_{t-1}}). \tag{2.3}$$

The West Virginia model has been used to project the output of its 48 sectors to 1975. It also provides the information needed to estimate investment requirements to achieve these output levels. Since a matrix of labor coefficients is available—although not part of the basic model described above—it is possible to estimate the employment by sector that will be associated with projected output levels. All of the projections were made before air pollution abatement standards were adopted. Thus they represent projected output levels that are not constrained in any way by the cost of pollution abatement.[12]

Cost Adjustments Due to Pollution Abatement

As Leontief and Ford have pointed out, the empirical implementation of an input-output pollution abatement model "requires detailed knowledge of the input structure of such anti-pollution or non-polluting processes. Such information is, in particular, indispensable for all kinds of cost and price computations related to a choice of appropriate levels of anti-pollution activities."[13] The requisite data were not available at the time Leontief and Ford wrote their paper, and the results they published are based on what they consider to be "rather crude estimates of the wage and interest costs" of some typical control processes and their own estimates of price effects.

The major empirical objective of the research reported here was to collect data on the cost of installing and operating pollution abatement equipment for the major industrial air polluters located in West Virginia. The details about data collection are given in later chapters. In terms of the present discussion, the objective was to collect primary data that would permit us to adjust the elements of the basic matrices in the input-output model to reflect the added costs incurred by the region's major polluters as they comply with prescribed emission standards. The direct and indirect costs of compliance were estimated by a two-step procedure. First, the following system was solved:

$$\mathbf{X}'_t = (\mathbf{I} - \mathbf{A}'_t - \mathbf{D}' - \mathbf{B}')^{-1}(\mathbf{Y}_t - \mathbf{B}'\mathbf{X}_{t-1}). \tag{2.4}$$

The primes in (2.4) indicate that some elements in each of the matrices have been changed to reflect the capital and operating costs of pollution abatement. The final "real" costs incurred by each vector were then obtained by solving the following:

$$\mathbf{X}^*_t = (\mathbf{X}'_t - \mathbf{Y}_t) - (\mathbf{X}_t - \mathbf{Y}_t), \tag{2.5}$$

in which a constant vector of final demand has been subtracted from the two output vectors. Since a constant level of final demand was specified for both the unconstrained and constrained projections, it was necessary to remove this constant term from both projections to accurately measure the relative change in inter-industry costs due to pollution abatement. $\mathbf{X}^*_t$ thus yields the estimated incremental production that will be required to produce a fixed bill of goods when all of the establishments in the state have complied with existing pollution abatement requirements. This is what was referred to above as our estimate of the "real" cost of pollution abatement. The elements of this vector show the distribution of the additional production requirements by sector.

The West Virginia Particulate Pollution Coefficients

During the early 1970s, we estimate that particulate emissions in West Virginia amounted to about 975,000 tons per year. Electric utilities accounted for more than half the total. The second largest industrial polluter, accounting for about 23 percent of total particulate emissions, was the primary metals sector. The chemical industry, producing almost 12 percent of West Virginia's particulate emissions, ranked third. Stone and clay products, which produced about 8 percent of total emissions, ranked fourth. Together, these 4 sectors accounted for 93 percent of the total output of particulates in West Virginia. Only 9 other sectors generated measurable amounts of particulate emissions.

Full compliance with the 1970 regulations would lead to an impressive drop of about 90 percent of airborne particulates. The largest declines would be experienced by the 4 heavy polluters.

Estimates of particulate emissions by sector and the 1965 West Virginia input-output tables were used to compute direct air pollution-output coefficients that are expressed in thousands of tons emitted per million dollars of total gross output. These coefficients were kept in 1965 prices to permit rough comparison with the national coefficients calculated by Leontief and Ford.[14] The West Virginia coefficients, given in Table 2-1, differ in most cases from the Leontief-Ford coefficients primarily because of differences in industrial classification. In general, the West Virginia coefficients are larger than the national coefficients. For example, the West Virginia electric utility coefficient of 1.4703 compares with the national coefficient of 0.27100. But virtually all electric energy in West Virginia is generated by coal-burning steam plants whereas the national coefficient includes the output of hydroelectric, nuclear, and other generating facilities that do not pollute the atmosphere. Other comparisons are more difficult to make because of the aggregation problem. Leontief and Ford show, for example, that the national lime industry has an unusually high coefficient of 2.16350. The West Virginia coefficient for stone and clay products (including lime products) is .55000. This is not because West Virginia's limestone operations are cleaner than the national average, but simply because all other stone and clay products—many of which contribute little or nothing to air pollution—are included in this sector in the West Virginia tables.

Table 2-1 also shows what the pollution-output coefficients would have been

Table 2-1
Direct Particulate Air Pollution–Output Coefficients for West Virginia (Thousands of Tons Emitted per Million Dollars of Output in 1965 Prices)

		(1) Before Abatement	(2) After Abatement a/	(3) Absolute Change	(4) Per Cent Change
2.	Coal mines (undgd.)				
3.	Coal (strip & auger)	.03471	.01619	-.01852	-53.4%
4.	Petroleum & natural gas	.00264	.00264	0	0
17.	Chemicals	.09500	.00646	-.00854	-93.2
18.	Petroleum products	.08930	.00670	-.08260	-92.5
19.	Glass	.00334	.00179	-.00155	-46.4
20.	Stone & clay products	.55000	.04035	-.50965	-92.7
21.	Primary metals	.27000	.01092	-.25908	-96.0
41.	All other services	.00200	.00200	0	0
42.	Railroads	.00164	.00164	0	0
43.	Trucking & warehousing	.13870	.13870	0	0
44.	All other transportation	.01300	.01300	0	0
46.	Electrical systems	1.47030	.09687	-1.37343	-93.4
48.	Water & sanitary services	.03300	.01597	-.01703	-51.6
52.	Total	.05210	.00520	-.04697	-90.0%

a/ To meet primary standards.

if 1970 primary air quality standards had been met. As suggested by the earlier discussion of total emissions, the largest reductions would be in the coefficients of the heavy polluters. There would be a comparable drop in the coefficient for petroleum products, a relatively small sector in West Virginia. There would be no change in the coefficients of five sectors with measurable but not significant levels of particulate pollution. The coefficients of the remaining sectors would be reduced by about one-half.

Only 13 industrial sectors in West Virginia contribute directly to particulate pollution. But these sectors purchase some of their inputs from other West Virginia sectors. Thus when the amount of particulate pollution generated directly and indirectly is taken into account, the contribution that every sector makes to total air pollution can be estimated.[15] About 66 percent of particulate pollution in West Virginia can be directly attributed to the major polluters listed in Table 2-1. The remaining one-third is generated indirectly by suppliers to the major polluters (who in many cases are also customers of the major polluters). Further details by industrial sector are given in Table 2-2, which shows the contribution of each sector to West Virginia's total particulate pollution.[16]

The West Virginia Sulfur Dioxide Pollution Coefficients

Annual emissions of sulfur dioxide in West Virginia during the early 1970s amounted to an estimated 1.3 million tons. The major contributors were electric utilities, primary metals, and chemicals. Twenty-four other sectors contributed directly to sulfur dioxide pollution, but by substantially smaller amounts.

In many cases sulfur dioxide and particulates are emitted together, as in the case of steam-generating power plants burning high-sulfur coal. And the abatement of both types of emissions is generally undertaken simultaneously.[17] It is useful, however, to separate the two major pollutants (particulates and sulfur dioxide) for analytical purposes. As indicated earlier, the separate abatement cost estimates in this and the following chapters are combined in Chapter 5.

The direct sulfur dioxide pollution-output coefficients for West Virginia are given in Table 2-3. The estimated reductions in the direct coefficients of the three major polluters that will be required to meet the primary standards established by 1970 legislation are also given. The West Virginia sulfur dioxide coefficients for electric utilities and primary metals are substantially larger than the national coefficients estimated by Leontief and Ford.[18] It is difficult to make a comparison in the case of chemicals since there has been much more disaggregation of the chemical industry in the national input-output table than in the West Virginia table. As in the case of the particulate coefficients, the differences between West Virginia and the nation as a whole are due partly to

Table 2-2

Direct and Indirect Particulate Pollution–Output Coefficients for West Virginia (Thousands of Tons of Emissions Directly and Indirectly Generated per Million Dollars of Sales to Final Demand in 1965 Prices)

	(1) Before Abatement	(2) After Abatement a/	(3) Absolute Change	(4) Per Cent Change
1. Agriculture	.02714	.01675	-.01039	-38.3%
2. Coal mines (undgd.)	.08896	.05506	-.03390	-38.1
3. Coal (strip & auger)	.06954	.09392	.02438	35.1
4. Petroleum & natural gas	.02764	.01657	-.01107	-40.0
5. All other mining	.03723	.02571	-.01152	-30.9
6. Building contractors	.09169	.02316	-.06853	-74.2
7. Non-building contractors	.08188	.02181	-.06007	-73.4
8. Special contractors	.01994	.00994	-.01000	-50.2
9. Food products (n.e.c.)	.04854	.03577	-.01277	-26.3
10. Dairies	.01842	.01269	-.00573	-31.1
11. Bakeries	.08644	.06430	-.02214	-25.6
12. Beverages	.01819	.01200	-.00619	-34.0
13. Apparel & accessories	.01525	.01168	-.00357	-23.4
14. Logging & sawmills	.02715	.01994	-.00721	-25.6
15. Furn. & wood fabrication	.01859	.01290	-.00569	-30.6
16. Printing & publishing	.01206	.00945	-.00261	-21.6
17. Chemicals	.13922	.02688	-.11234	-80.7
18. Petroleum products	.12560	.02585	-.09975	-79.4
19. Glass	.07436	.05191	-.02245	-30.2
20. Stone & clay products	.63000	.08905	-.54095	-85.
21. Primary metals	.28757	.02492	-.26265	-91.3
22. Fabricated metals	.05089	.01562	-.03527	-69.3%

Table 2-2 (cont.)

	(1) Before Abatement	(2) After Abatement	(3) Absolute Change	(4) Per Cent Change
23. Machinery (except electric)	.02740	.01592	-.01148	-41.9%
24. Electric machinery	.10374	.02829	-.07545	-72.7
25. Transportation equipment	.03556	.01769	-.01787	-50.2
26. Instruments & products	.01274	.01047	-.00227	-17.8
27. All other manufacturing	.01684	.01214	-.00470	-27.9
28. Restaurants & bars	.08068	.06006	-.02062	-25.6
29. Wholesale trade	.03637	.02750	-.00887	-23.9
30. Retail food stores	.05355	.03979	-.01376	-25.7
31. Auto service stations	.02860	.02130	-.00730	-25.5
32. All other retail	.03329	.02467	-.00862	-25.9
33. Banking	.01598	.01175	-.00423	-26.5
34. Other finance	.00714	.00609	-.00105	-14.7
35. Insurance agents	.00251	.00184	-.00067	-26.7
36. Real estate	.02147	.01221	-.00926	-43.1
37. All other FIRE	.01123	.00859	-.00264	-23.5
38. Hotels & lodgings	.09183	.06826	-.02357	-25.7
39. Medical & legal services	.01176	.00837	-.00339	-28.8
40. Educational services	.03146	.01838	-.01308	-41.6
41. All other services	.02888	.01820	-.01068	-37.0
42. Railroads	.01380	.00912	-.00468	-33.9
43. Trucking & warehousing	.15330	.14850	-.00480	- 3.1
44. All other transportation	.02884	.02292	-.00592	-20.5
45. Communications	.01597	.01179	-.00418	-26.2
46. Electrical systems	1.57509	1.17477	-.40032	-25.4
47. Gas systems	.00774	.00486	-.00288	-37.2
48. Water & sanitary services	.13647	.08703	-.49440	-36.2%

a/ To meet primary standards.

Table 2-3

Direct Sulfur Dioxide Pollution–Output Coefficients for West Virginia (Thousands of Tons Emitted per Million Dollars of Output in 1965 Prices)

	(1) Before Abatement	(2) After Abatement	(3) Absolute Change	(4) Per Cent Change
1. Agriculture	.00749	.00749	0	0%
4. Petroleum & natural gas	.06420	.06420	0	0
6. Building contractors	.01361	.01361	0	0
9. Food products (n.e.c.)	.00430	.00430	0	0
10. Dairies	.00430	.00430	0	0
11. Bakeries	.00430	.00430	0	0
12. Bevarages	.00430	.00430	0	0
14. Logging & sawmills	.00690	.00690	0	0
16. Printing & publishing	.00330	.00330	0	0
17. Chemicals	.05000	.03180	-.01820	-36.4
20. Stone & clay products	.01300	.01300	0	0
21. Primary metals	.16500	.06630	-.09870	-59.8
22. Fabricated metals	.05940	.05940	0	0
23. Machinery (except electric)	.01000	.01000	0	0
24. Electric machinery	.00830	.00830	0	0
25. Transportation equipment	.00130	.00130	0	0
30. Retail food stores	.00770	.00770	0	0
31. Auto service stations	.00770	.00770	0	0
32. All other retail	.00770	.00770	0	0
34. Other finance	.00340	.00340	0	0
35. Insurance agents	.00340	.00340	0	0
41. All other services	.00960	.00960	0	0
42. Railroads	.00660	.00660	0	0
43. Trucking & warehousing	.00273	.00273	0	0
44. All other transportation	.03900	.03900	0	0
45. Communications	.00760	.00760	0	0
46. Electrical systems	2.20000	1.78000	-.42000	-19.1%

aggregation. But we expected a substantially larger coefficient for electric utilities, since virtually all electric energy generated in this state comes from steam-generating plants, some of which were burning relatively high-sulfur coal in the early 1970s. For the same reason we expected a somewhat larger coefficient for the primary metals sector.

In relative terms, the changes required to meet the primary air quality standards are not as large for sulfur dioxide as those discussed earlier for particulates. The coefficient for primary metals would have to be reduced by almost two-thirds, and this must be considered a relatively large change. But the electric utility coefficient would decline by less than 20 percent to meet primary standards. This relatively modest decline is explained in part by the location of some of the state's power plants. Those located in Priority III regions will not be required to take action to reduce sulfur dioxide emissions since they are now within control limits. The coefficients in Table 2-3 are statewide averages, and the changes shown are not at all indicative of the magnitude of changes that will have to be made at individual plants.

The direct and indirect sulfur dioxide-output coefficients are given in Table 2-4. This is the counterpart of Table 2-2 for particulates and requires little additional comment. It shows that while the three major polluters are the primary contributors of sulfur dioxide, all sectors of the West Virginia economy contribute at least indirectly to the total volume of annual emissions. Our estimates of the direct and indirect coefficients that would satisfy primary standards are given in Column 2 of this table. The last two columns show the absolute and relative changes in the direct and indirect coefficients that will be required to meet these standards.

West Virginia Hydrocarbon and Carbon Monoxide Pollution–Output Coefficients

We were able to collect data on hydrocarbon and carbon monoxide emissions from stationary sources for only 7 sectors.[19] Relatively little will be said about these "minor" pollutants. Comparison with earlier tables shows that the volume of hydrocarbons and carbon monoxide emitted in West Virginia is not large. The data presented in Table 2-5 have been included primarily for the sake of completeness. Nothing further will be said about the control of these two minor pollutants in this study. We have already indicated, however, that the reduction of particulate emissions will in many cases lead to a reduction of hydrocarbon emissions. Again, for the sake of completeness, we calculated the direct and indirect hydrocarbon- and carbon monoxide-output coefficients. These are presented in Table 2-6.

Table 2-4
Direct and Indirect Sulfur Dioxide Pollution–Output Coefficients for West Virginia (Thousands of Tons of Emissions Directly and Indirectly Generated per Million Dollars of Sales to Final Demand in 1965 Prices)

		(1) Before Abatement	(2) After Abatement a/	(3) Absolute Change	(4) Per Cent Change
1.	Agriculture	.03998	.03393	-.00605	-15.1%
2.	Coal mines (undgd.)	.08255	.06682	-.01573	-19.1
3.	Coal (strip & auger)	.03237	.02632	-.00605	-18.7
4.	Petroleum & natural gas	.09348	.08864	-.00484	-5.2
5.	All other mining	.04763	.03835	-.00928	-19.5
6.	Building contractors	.05406	.04446	-.00960	-17.8
7.	Non-building contractors	.03122	.02450	-.00672	-21.5
8.	Special contractors	.01988	.01460	-.00528	-26.6
9.	Food products (n.e.c.)	.07599	.06273	-.01326	-17.4
10.	Dairies	.03032	.02593	-.00439	-14.5
11.	Bakeries	.13217	.10797	-.02420	-18.3
12.	Beverages	.02449	.02010	-.00439	-17.9
13.	Apparel & accessories	.01968	.01598	-.00370	-18.8
14.	Logging & sawmills	.04556	.03873	-.00683	-15.0
15.	Furn. & wood fabrication	.02500	.02032	-.00468	-18.7
16.	Printing & publishing	.01420	.01223	-.00197	-13.9
17.	Chemicals	.09269	.06439	-.02830	-30.5
18.	Petroleum products	.03824	.03149	-.00675	-17.6
19.	Glass	.10033	.08106	-.01927	-19.2
20.	Stone & clay products	.09932	.08276	-.01656	-16.7
21.	Primary metals	.18705	.08393	-.10312	-55.1
22.	Fabricated metals	.05249	.03571	-.01678	-32.0
23.	Machinery (except electric)	.03748	.02946	-.00802	-21.4
24.	Electric machinery	.12084	.08835	-.03249	-26.9%

Table 2-4 (cont.)

		(1) Before Abatement	(2) After Abatement	(3) Absolute Change	(4) Per Cent Change
25.	Transportation equipment	.03648	.02494	-.01154	-31.6%
26.	Instruments & products	.01086	.00817	-.00269	-24.8
27.	All other manufacturing	.02209	.01766	-.00443	-20.0
28.	Restaurants & bars	.11985	.09721	-.02264	-18.9
29.	Wholesale trade	.05020	.04081	-.00939	-18.7
30.	Retail food stores	.08540	.07071	-.01469	-17.2
31.	Auto service stations	.04425	.03754	-.00671	-15.2
32.	All other retail	.05375	.04517	-.00858	-16.0
33.	Banking	.02329	.01894	-.00435	-18.7
34.	Other finance	.01372	.01180	-.00192	-14.0
35.	Insurance agents	.00695	.00630	-.00065	-9.5
36.	Real estate	.02478	.02021	-.00457	-18.4
37.	All other FIRE	.01527	.01246	-.00281	-18.4
38.	Hotels & lodgings	.13313	.10794	-.02519	-18.9
39.	Medical & legal services	.01613	.01316	-.00297	-18.4
40.	Educational services	.03616	.02943	-.00673	-18.6
41.	All other services	.03696	.03178	-.00518	-14.0
42.	Railroads	.02023	.01731	-.00292	-14.4
43.	Trucking & warehousing	.02161	.01798	-.00363	-16.8
44.	All other transportation	.06289	.05894	-.00395	-6.3
45.	Communications	.02618	.02269	-.00349	-13.3
46.	Electrical systems	2.34144	1.89445	-.44699	-19.1
47.	Gas systems	.01493	.01355	-.00138	-9.2
48.	Water & sanitary services	.13657	.11010	-.02647	-19.4%

a/ To meet primary standards.

Table 2-5

Direct Hydrocarbon and Carbon Monoxide Pollution–Output Coefficients for West Virginia (Thousands of Tons Emitted per Million Dollars of Output in 1965 Prices)

	HC	CO
4. Petroleum & natural gas	.02580	.05529
17. Chemicals	.00830	.01800
42. Railroads	.02020	.06700
43. Trucking & warehousing	.01150	.00544
44. All other transportation	.02960	.94030
46. Electrical systems	.00870	.03000
48. Water & sanitary services	.28800	.56000

Table 2-6
Direct and Indirect Hydrocarbon and Carbon Monoxide Pollution–Output Coefficients for West Virginia (Thousands of Tons Emitted Directly and Indirectly per Million Dollars of Sales to Final Demand in 1965 Prices)

	HC	CO
1. Agriculture	.00038	.00046
2. Coal mines (undgd.)	.00043	.00026
3. Coal (strip & auger)	.00054	.00067
4. Petroleum & natural gas	.02771	.05899
5. All other mining	.00071	.00081
6. Building contractors	.01207	.02309
7. Non-building contractors	.00169	.00268
8. Special contractors	.00058	.00083
9. Food products (n.e.c.)	.00103	.00148
10. Dairies	.00072	.00115
11. Bakeries	.00135	.00177
12. Beverages	.00192	.00333
13. Apparel & accessories	.00052	.00067
14. Logging & sawmills	.00088	.00123
15. Furn. & wood fabrication	.00061	.00098
16. Printing & publishing	.00082	.00099
17. Chemicals	.01108	.02325
18. Petroleum products	.00352	.00745
19. Glass	.00080	.00084
20. Stone & clay products	.00147	.00171
21. Primary metals	.00092	.00159
22. Fabricated metals	.00104	.00182
23. Machinery (except electric)	.00073	.00126
24. Electric machinery	.00325	.00666
25. Transportation equipment	.00121	.00153
26. Instruments & products	.00684	.00162
27. All other manufacturing	.00045	.00049
28. Restaurants & bars	.00115	.00143
29. Wholesale trade	.00130	.00229
30. Retail food stores	.00155	.00243
31. Auto service stations	.00179	.00297
32. All other retail	.00164	.00272
33. Banking	.00044	.00071
34. Other finance	.00028	.00059
35. Insurance agents	.00018	.00034
36. Real estate	.00231	.00454
37. All other FIRE	.00103	.00190
38. Hotels & lodgings	.00216	.00329
39. Medical & legal services	.00214	.00400
40. Educational services	.00201	.00364
41. All other services	.00141	.00249
42. Railroads	.02100	.00723
43. Trucking & warehousing	.01155	.00613
44. All other transportation	.03284	.10244
45. Communications	.00063	.00041
46. Electrical systems	.01005	.00407
47. Gas systems	.00350	.00769
48. Water & sanitary services	.29400	.57044

3 The Control of Particulate Emissions

All of the major manufacturing industries in West Virginia will be affected to some extent by the particulate emission standards set by the West Virginia Air Pollution Control Commission (WVAPCC). Emission standards are more stringent for large plants (or establishments) than for smaller ones. Thus, large companies will bear what might appear to be a disproportionate share of the cost of particulate control. But in some small, marginal plants—particularly those burdened with outmoded equipment and poor productivity records—the cost of meeting particulate standards may be more than the firm can bear. These plants will close. Some of them would, no doubt, be driven out of existence by even less stringent and costly environmental regulations. While the economic impact of particulate regulations will vary widely from company to company in West Virginia, the overall impact is not likely to be economically significant. Our judgment is that economic development in West Virginia will not be retarded by strict enforcement of existing particulate regulations.

The Effect of Particulate Regulations on Specific Industries

In this study we are concerned only with the abatement of point, stationary particulate emissions. Such sources as burning gob and refuse piles will not be considered. We will, however, discuss the steps that must be taken to meet primary and secondary air quality standards. The relevant regulations established by the WVAPCC for particulate control include those that will prevent and control particulate air pollution from:

1. Combustion of fuel in indirect heat exchanges (Regulation II).
2. The operation of hot-mix asphalt plants (Regulation III).
3. The operation of coal preparation plants and coal-handling operations (Regulation IV).
4. The prevention and control of particulate air pollution from manufacturing process operations (Regulation VII).

Regulations II and VII are the most inclusive and will affect the largest number of West Virginia establishments. These regulations include emission limits on both total pounds per hour and on the opacity of the exit gas streams.

The pounds-per-hour emission limits were established to deal primarily with the large particles that contribute the bulk of weight to particulate emissions. Opacity regulations are intended to deal with the far more numerous but very minute particles that reflect light more than the larger particles. These small particles lead to the coloring of smoke and haze and to decreased visibility. Opacity limits are important not only for aesthetic reasons but also because of health considerations. In general, the smaller particles have the greatest carcinogenic effect. Certain hazardous chemicals are also controlled by particulate regulations, as these chemicals are emitted in the form of acid mists; and the latter are classified as particulates for regulatory purposes.

Some specific examples of industry problems are given in the following paragraphs. The numbers in parentheses following the names of the industries are the sector numbers in the West Virginia input-output model.

Coal Mines (2 and 3). Particulate emissions caused by the coal industry result largely from the drying of crushed coal that has been put through one of the various washing processes used to reduce the volume of pyrites in the coal. Health and explosion hazards from dust emissions in the mines themselves are covered by the Occupational Health and Safety Act of 1970 rather than by the Clean Air Act.

Petroleum Products (4). The boilers at a petroleum refining complex can create a particulate problem if the fuel burned is coal. Other problems include hydrocarbon emissions from storage tanks and pipes and the entrapment in exit gases of fine particles used as heat-transfer and chemical-reaction media in fluidized-bed cracking units. There is a similar problem of carbon carry-over in briquet plants, but this leads to smaller emissions than those from carbon manufacturing sites. The carry-over of tar particles in asphalt plants creates specific problems that are subject to Regulation III. The problem is localized, however, since asphalt plants tend to be relatively small, and the particulates they emit are not widely dispersed.

Chemicals (17). The most difficult and costly emissions to control are the ash particles from combustion in "indirect-fired heat exchangers"—the industrial boilers used to supply steam in chemical manufacturing processes. The emissions are costly to control because of the large amounts of coal used. The combustion of coal is, of course, the cause of subsequent ash emissions. This type of emission is sometimes overlooked in discussions of air pollution caused by the chemical industry, which frequently center on the chemical processes themselves.

There is also a problem of controlling noxious odors in the vicinity of chemical plants. Even before the establishment of air quality standards, however, the chemical industry attempted to limit these odors because of the economic necessity of recovering some of the chemicals causing odors in exit gases. Compliance with published standards will lead to a further reduction in odors and particulate emissions since the most stringent mass emission limits are those that apply to chemical processes. In particular, acid mists, NH_3, carbon emissions (from carbon manufacturing processes), emissions from calcining (drying) operations, and carry-over residues from polymer manufacturing must all be rigidly controlled under present standards.

Glass (19). The fine particles carried out of the stacks of glass furnaces can be a major health hazard. Although the weight of these emissions is relatively small, present standards require that they be severely limited. The drying or calcining of clay may also produce particulate emissions caused by air swirls around the clay. Particles are created in glass factories from sand as it is melted to fuse the glass ingredients under intense heat. Glass furnaces are typically fired by natural gas, and the combustion of this gas itself creates no particulate problem.

Primary Metal Products (21). Among the prime sources of emissions from steel and other primary metal plants are large industrial boilers and furnaces. There are also process emissions from stack and furnace openings in the cupolas of gray-iron foundries, brass factories, steel-mill blast furnaces, steel-cintering operations, and the electric-arc and induction furnaces used in specialty steel mills. Particulates from these sources are to be controlled under the mass emission provision of the Clean Air Act. Coke-producing plants are included in the primary metals sector, and their emissions are to be controlled primarily under the opacity restrictions of the law.

The major problems in by-product ovens are leaks around doors and covers and the creation of hot, smoky clouds when coal is charged to an oven or when hot coke is pushed from the oven into transit cars. The by-product ovens collect many of the gases emitted when coal is pyrolized (heated in the absence of air). Beehive coke ovens have historically created more dust and gaseous hydrocarbon emissions than most other industrial processes and, in the absence of controls, would be major air polluters.

Stone and Clay Products (20). Digging and transporting stone and clay materials are the two major activities causing fugitive dust. Dust also escapes around conveyors and the machinery used in crushing operations. Since lime is manufactured by heating limestone–by the combustion of gas, oil, or coal directly over the stone–the carry-over of dust, and ash from coal combustion, may be other sources of particulate pollution in this sector. Cement plants have similar problems, requiring similar solutions, to those of the limestone and limestone products industry.

Fabricated Metal Products and "Other Manufacturing" Industries (22 and 27). A wide variety of process operations can cause fine-particle emissions. Such processes include shaping, polishing, and similar operations. In general, the processing of most solid products causes localized particulate emission problems. The particulate pollution problems created by these industries are relatively small when compared with those of the major air polluters. There are some highly specialized problems, such as mist emissions from paper mills, or the relatively small amounts of hydrocarbon emissions from gas processing plants. Some of these processes also produce highly noxious odors. But because the volume of emissions is relatively small and the emissions tend to be quite localized, the particulate pollution problems caused by the wide range of industries included in this catch-all category are relatively minor.

Electric Utilities and Systems (46). There are more stringent limits on allowable emissions from the stacks of electric utility boilers than from industrial boilers. Utility emissions of ash are limited to .05 pounds per million BTUs of heat input. The comparable standard for industrial boilers is .09 pounds per million BTUs. Both of these limits are more stringent than those required by federal regulations on new fuel-burning units. The latter have a limit of .10 pounds per million BTUs. All of the utility ash emissions in West Virginia result from the combustion of coal, since coal-burning plants account for 99 percent of the power generated in the Mountain State.

Methods of Control

There is usually more than one method available to control particulate emissions from a given source and process. The method chosen will depend on its cost and efficiency. Large enterprises, in particular, must devise an overall strategy of control. In a small establishment only one or two pieces of equipment may be needed, although the cost of these may be high relative to the net worth of the establishment. But in large industrial operations, such as steel mills and chemical complexes, particulate emissions can be controlled only if an overall strategy is designed.

One element of a particulate control strategy includes the installation, where necessary, of specific pieces of equipment. Listed roughly in the order of their effectiveness and importance these are: (1) electrostatic precipitators, (2) baghouses, (3) wet collection devices (scrubbers), and (4) mechanical collectors such as cyclones, settling chambers, and similar devices. Another element of a control strategy is the redesign of equipment to reduce emissions. This could involve major capital expenditures, and this strategy is likely to be pursued only if old equipment has been fully depreciated and can be scrapped. Finally, particulate emissions from combustion can be reduced by fuel-switching, such as from coal to gas. This strategy can be followed, however, only if a cleaner fuel is available at a competitive price.

In certain processes, the major problem is that of fugitive dust control. Essentially, this consists of "plugging leaks" to contain particulates within the plant for ultimate disposal in ways that will not add to air pollution. In designing plants to comply with existing air quality standards, it is essential that efficient and leak-proof ductwork be used to collect particulate-bearing gases for transfer to the control devices that eliminate the particulates.

Electrostatic precipitators (see Figure 3-1) are highly efficient devices for the control of ash from coal combustion. They are used extensively to control ash emissions from utility and industrial boilers, as well as emissions from lime and cement kilns. Precipitators can be used to some extent to control the emission of small particles, but they do not work well in corrosive and high temperature environments. They also do not perform as effectively as might be desired in processes that use materials with more than a small amount of electrical

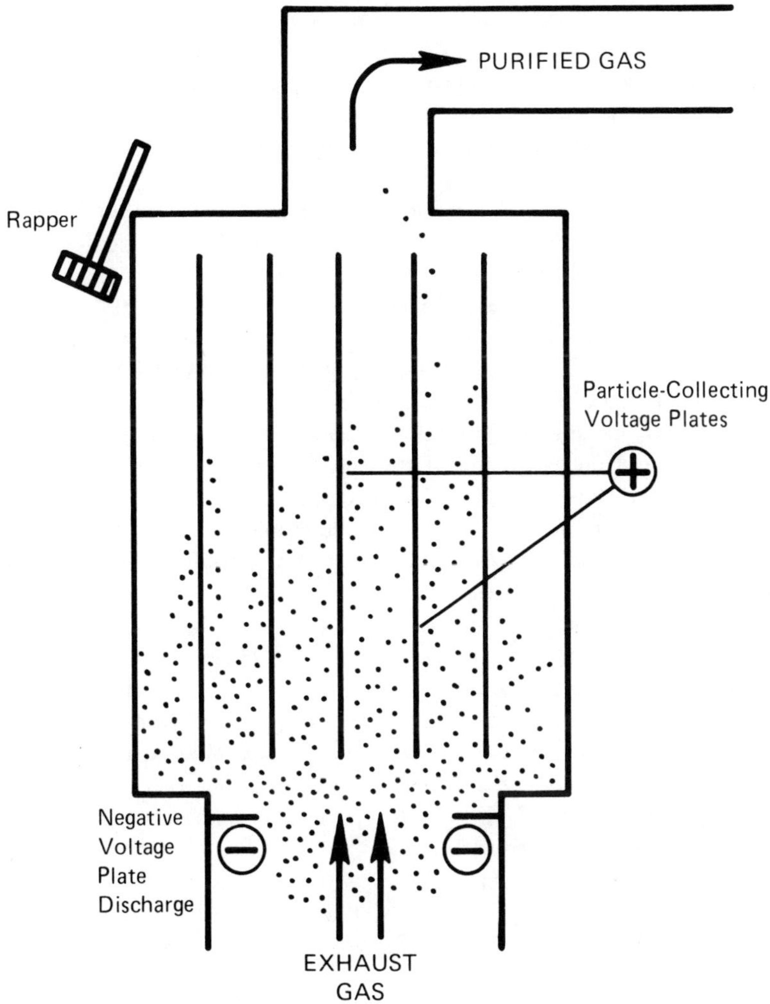

Figure 3-1. Electrostatic Precipitator

resistance. Thus, electrostatic precipitators are not used extensively outside of boiler stacks and calcining operations.

When electrostatic precipitators are attached to existing equipment, new ductwork must usually be installed to transport exit flue gases to the rather bulky equipment required to extract particles.[1] Maintenance costs on electrostatic precipitators are generally low, however, as there are few moving parts. Electrical plates and connections are easily maintained except under the extreme conditions mentioned in the preceding paragraph.

Precipitators work by imparting an electrical charge from voltage-plate electrodes to small particles by corona discharge as the gas containing the particles passes by. In an electrical field the charge causes the individual particles to drift to the wall and stick in conglomerates of particles. Periodically, the walls are rapped to knock the conglomerates loose into a collection bin. Electrostatic precipitators effectively control small particles, in the range of 5 to 30 micrometers, to removal efficiencies of more than 99 percent.

Baghouses (see Figure 3-2) are used to collect dry particles that may vary widely in size. A cloth bag filters particles while allowing gases to pass on to the atmosphere. Periodic shaking frees the particles to fall into bins. Only selected fabrics can be used for a given application, if continuous operations are to be maintained without fabric failure or caking and clogging of the cloth. The major costs associated with the construction of baghouses involve ductwork housing and air fans. The cost of the bags themselves is rarely a major item. By placing hoods over open metal-making furnaces, baghouses can be used to collect and eliminate particles from gas streams. Extremely fine particles cannot be collected in bags, however. Thus, in addition to their use on certain types of furnaces, baghouses most often are constructed in conjunction with stone and limestone facilities, certain calcining (drying) operations, and in conjunction with various physical manufacturing processes.

There are three types of wet collection devices (see Figure 3-3) that can be used for the abatement of particulate pollution. These are: (a) *wet packed towers*, which are tanks or towers containing solid pellets with a downward flow of water through which the gas stream is passed; (b) *impingement and cyclone scrubbers*, in which the gas stream is forced into circular flows containing sprays of water; and (c) *venturi and other scrubbers*, consisting of ducts that force particles into jets of water. In venturi-type scrubbers the duct is tapered to increase the rate at which the gas flows.

Scrubbers work by wetting the particles, which has the effect of making them heavier. This results in impingement of the wet particles against collecting surfaces. The collected particles are removed from the surfaces with a liquid flush.

Both the cost and efficiency of particulate collection increase as we move from tanks of water to water sprays, and thence to "high-energy" jets. The high removal standards required by the WVAPCC regulations generally dictate the use

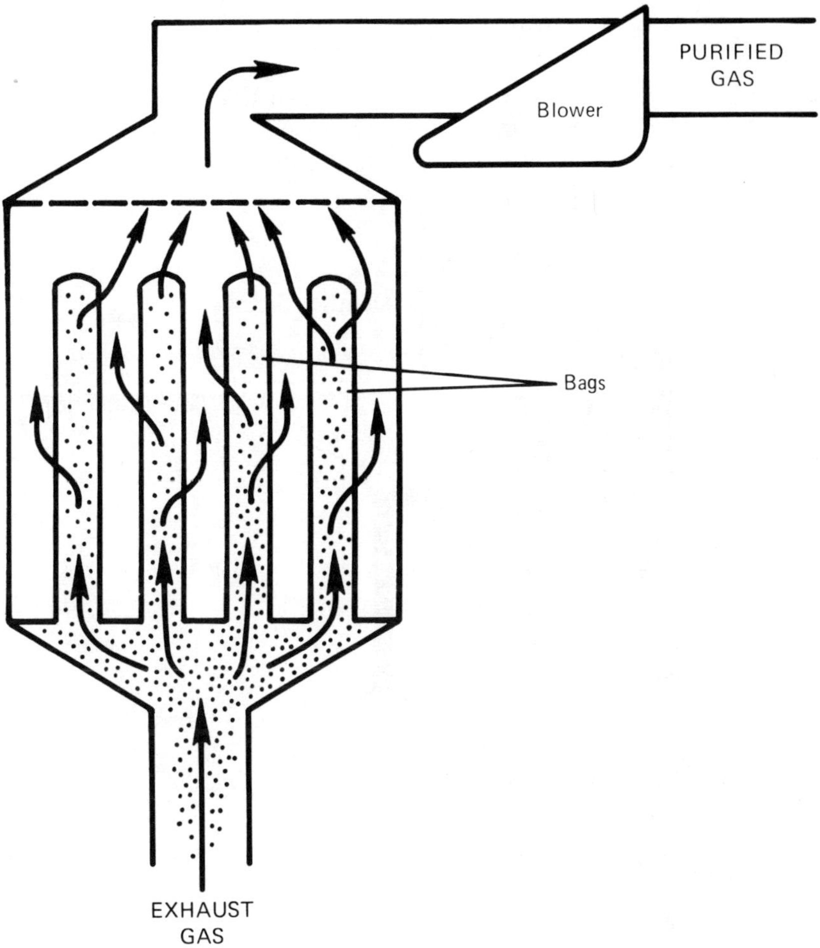

Figure 3-2. Baghouse Schematic

of high-energy, high-efficiency scrubbers. And because of the high recycle rate of water required, these devices tend to have high annual operating costs. They have the advantage, however, of being effective on all types of particles. In some cases they can also remove various chemicals, vapors, and gases, including SO_2, in addition to particulate matter.

Scrubbing devices are extensively used in the chemical industry; they are the primary means used to remove tar at asphalt plants, for example. They are also used to help control the effluents from blast and basic oxygen furnaces, from some industrial boilers, and to control coke-oven gas.

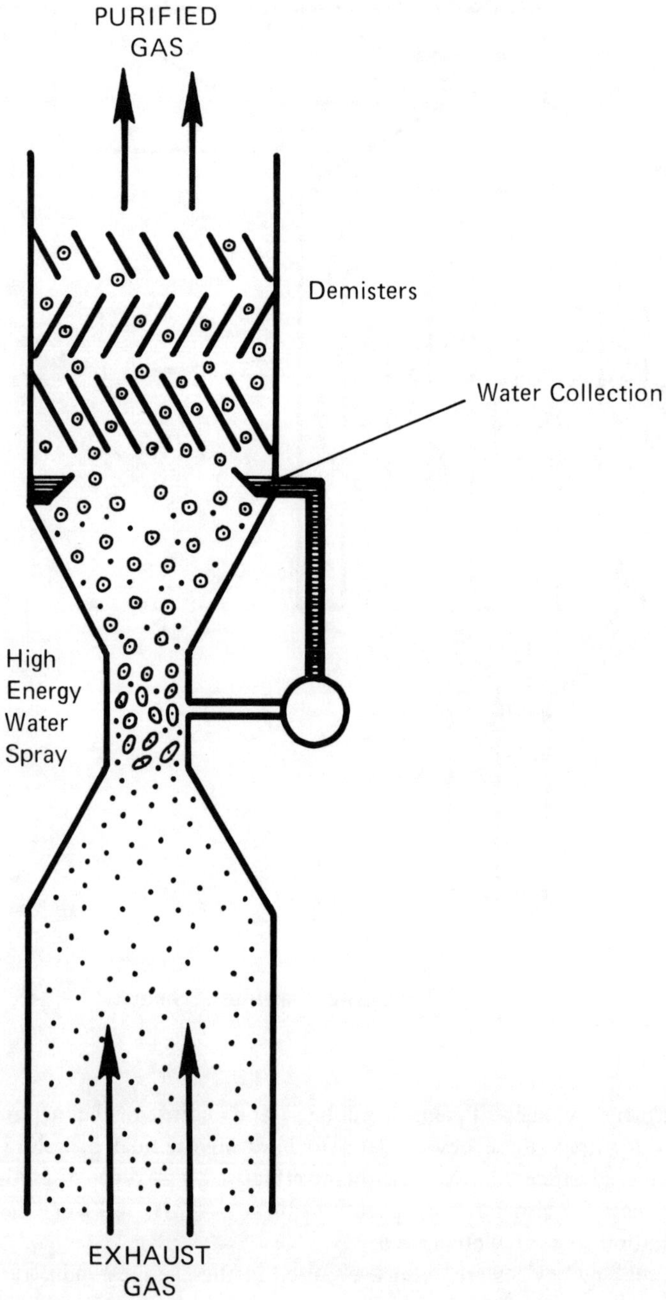

Figure 3-3. High-Energy Venturi Wet Scrubber

Mechanical collectors, such as cyclones (see Figure 3-4) or settling chambers are used to control emissions of large particles (greater than 20 micrometers). In general, they are not extremely efficient; but since they have no moving parts, their maintenance costs are negligible. Where mechanical collectors are used, gas flows are set in such a way that dry particles are directed to the walls or bottom of the collecting chamber. Given present emission control standards, mechanical collectors, generally, will be used by small industrial operations. They may be used by larger establishments, however, to remove the heavy particles from gas streams, which reduces or prevents clogging in subsequent collection devices such as the scrubbers described above.

Redesign of equipment, including the addition of hoods, is often necessary to eliminate the conditions that create particulate emissions. This is particularly true where open access allows particulate-laden gases to escape without easy means of channeling the gases. Redesign will almost certainly be necessary if coke plants in West Virginia are to meet present air pollution standards. Many industrial cupolas, furnaces, and boilers will require redesign to satisfy the law as it now stands.

Fugitive dust control is maintained by wetting the grounds where dust

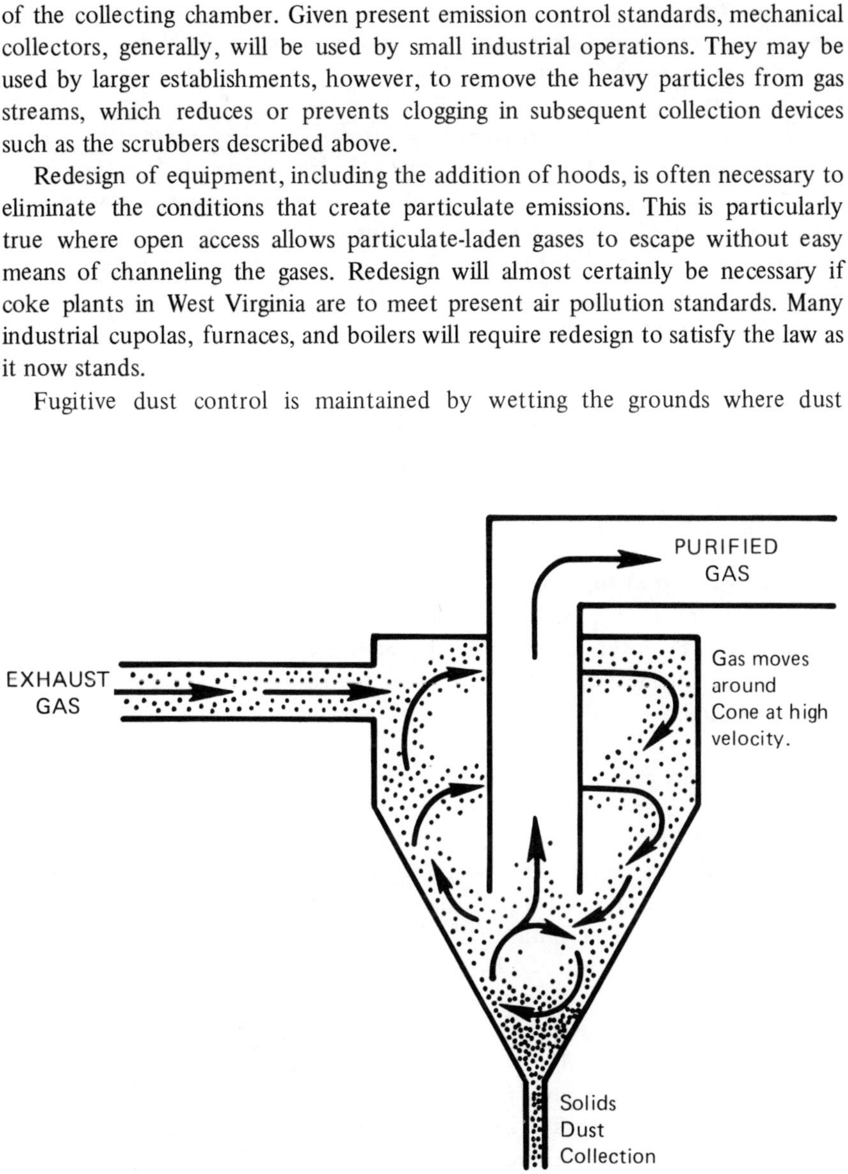

Figure 3-4. Cyclone Mechanical Collector

originates and by providing hoods and ductwork at certain solids-transfer points. In some cases fugitive dust control might require the restriction of operations. The control of fugitive dust may be a major problem in the stone and clay products sector where many establishments handle hundreds of tons of solid pellets daily.

At the level of the individual plant, particulate emission standards might be met by fuel-switching. This might consist of a switch from high-ash to a lower-ash coal or from coal to fuel oil or electric power. Clearly, however, every establishment cannot solve its air pollution problems by fuel-switching. Industrial oil is increasingly in short supply; and if every establishment attempted to solve its pollution problem by switching to electric energy, the electric energy sector would bear the entire direct burden of particulate abatement. Fuel-switching has been employed by a number of individual chemical and metal-producing firms in West Virginia. This has been true in particular where by switching fuels the firm feels it has a better chance of meeting both particulate and SO_2 emission standards.

Estimating the Costs of Particulate Control

The cost of controlling particulate emissions in the major industrial sectors of West Virginia was estimated in a number of ways using data from a variety of sources. Several plants in each of the industries considered to be major polluters were visited by staff members. Cooperating companies provided much of the cost information used in this study.[2] The data obtained by direct interview was supplemented by information from the pollution control literature, and the published compliance schedules of specific industries. In particular, we relied heavily on published estimates of the total capital costs of meeting compliance schedules.[3]

The cost estimates presented in this book tend to be somewhat higher than those given in previous studies.[4] Much of this is due to the higher costs of retrofitting of control equipment compared with the installation of such equipment in new plants. In particular, the cost of installing electrostatic precipitators in existing power plants may be at least three times as high as some early estimates in studies conducted for the Environmental Protection Agency, and in other studies conducted by private environmental and ecology groups.[5]

It is also important to point out that the cost estimates in this report were based on capital and operating budgets relating to 1972 and 1973. It is likely that the delivered cost of much of this equipment will be greater than the amount budgeted, particularly if installation is delayed for three or four years. This could easily happen since some construction is scheduled for completion in 1975 or later. Finally, the technical reliability of some pieces of control equipment may be less than that required to meet established standards. On field visits, engineering members of the project's staff were shown equipment that was not operating as efficiently as had been anticipated. In some cases, for example, high temperatures had caused difficulties with corrosion, and there were electrical problems that had not been experienced with similar equipment used previously under other conditions.

Pollution abatement equipment requires careful maintenance if excessive down time is to be avoided, and it appears that such service is difficult to obtain from manufacturers. Some establishments have actually installed excess abatement capacity anticipating that specific pieces of control equipment will operate at less than rated efficiency levels. Thus, despite the fact that the cost estimates published in this report may appear to be high—relative to those published earlier—we feel they are conservative.

The installation of pollution abatement equipment *per se* will not increase an establishment's productivity. Indeed, if there is excessive down time because the equipment does not meet published technical standards, productivity could be adversely affected. We encountered a number of cases, however, where management had coupled its pollution abatement program to a program of general modernization. In these cases every effort was made to distinguish between the cost of installing abatement equipment and the general cost of modernization. Also, certain types of equipment were installed for the abatement of both particulate and sulfur emissions. In such cases, the cost of the equipment was charged entirely to the estimates of particulate control given in this chapter. This may impart a slight upward bias to our estimates of the cost of particulate control. At the same time, it will impart a downward bias to the estimates of sulfur emission control discussed in the next chapter.[6] In general, where cost estimates were available only in terms of a range, we tended to lean toward the lower limits of that range.

The Direct Costs of Particulate Abatement

We estimate that the direct cost of installing particulate abatement equipment to meet West Virginia's clean air standards will be about \$309 million (expressed in 1975 prices). The direct annual cost of operating this additional equipment, again expressed in 1975 prices, we estimate will be about \$37.8 million. The details, by sector, are given in Table 3-1.[7] Electric utilities, which account for more than half of the state's particulate emissions, will also spend more than half of the capital funds required to meet the clean air standards.[8] The primary metals sector, which ranks second in terms of capital expenditures, accounts for slightly more than 28 percent of the total. And the chemical sector, which will spend about 15 percent of the total, ranks third. Together, these three sectors will account for more than 95 percent of all capital spending on particulate abatement equipment.

Table 3-1
Direct Capital and Operating Costs of Particulate Abatement

	Capital Costs (thousands of dollars)			Annual Operating Costs (thousands of dollars)		
	(1) Year of Estimate	(2) Amount in Current Dollars	(3) Amount in[a] 1975 Dollars	(4) Year of Estimate	(5) Amount in Current Dollars	(6) Amount in[b] 1975 Dollars
2. Coal mines (undgd.)	1969	$ 1,275	$ 2,409	1969	$ 440	$ 532
9. Food products (n.e.c.)	1967	174	245	1967	38	49
10. Dairies	1967	173	244	1967	35	45
11. Bakeries	1967	130	183	1967	26	34
12. Beverages	1967	173	244	1967	35	45
17. Chemicals	1971	41,700	46,724	1971	12,500	13,837
18. Petroleum products	1972	1,000	112	1972	200	217
19. Glass	1972	5,000	560	1972	1,000	1,085
20. Stone & clay products	1972	3,210	3,597	1972	642	697
21. Primary metals	1972	78,500	87,958	1972	15,400	16,714
22. Fabricated metals	1972	3,500	392	1972	700	760
23. Machinery (except electric)	1967	1,500	2,112	1967	300	332
24. Electric machinery	1967	740	1,042	1967	148	164
25. Transportation equipment	1967	1,700	2,394	1967	340	377
26. Instruments & products	1967	140	197	1967	280	310
46. Electrical systems	1972	143,700	169,009	1969	2,150	2,589
52. Total		$282,615	$309,422		$34,234	$37,787

[a] The inflator for capital goods was projected from data in the Economic Report of the President, 1973, Table C-3, p. 196, GPDI, Total nonresidential fixed investment.

[b] The inflator for operating costs was projected from data in the Economic Report of the President, 1973, Table C-3, p. 196, Total personal consumption expenditures.

Electric utilities, primary metals, and chemicals also account for the bulk (87.5 percent) of annual operating costs. But the distribution of these costs is quite different than that of the capital expenditures. Electric utilities will account for 52 percent of capital spending, but only 6.9 percent of the annual operating costs. And while 28 percent of capital spending to eliminate particulate pollution will be made by the primary metals sector, this sector will account for 44 percent of the annual operating costs. Similarly, while the share of capital spending accounted for by the chemical sector was slightly more than 15 percent, chemicals will account for 36.6 percent of annual operating costs. These sectoral differences in capital spending and operating costs are largely the result of variations in the type of abatement equipment that will be installed. Electric utilities will rely primarily on electrostatic precipitators to control emissions. Since these have few moving parts, they require little maintenance and are relatively inexpensive to operate. The primary metals and chemicals sectors will have to use wet scrubbers—including high-energy scrubbers—to effectively control some of their emissions. Although technically efficient, scrubbers have relatively high operating costs.

The differences discussed above are reflected in the capital and technical coefficients that are derived from the cost data. The new *total* expansion capital coefficients for the West Virginia model, which show the effects of capital spending on particulate abatement plant and equipment, are summarized in Table 3-2. We have focused on the total expansion capital coefficients in the text to summarize the changes and to give some idea of their magnitude. For

Table 3-2
Changes in Total Expansion Capital Coefficients (B_j) Due to Particulate Abatement Investment

	Original B_{ij}	New B_{ij}	Difference	Per Cent Change
2. Coal mines (undgd.) 3. Coal (strip & auger)	.686568	.691722	.005154	.75%
9. Food products (n.e.c.)	.186959	.193019	.006060	3.24
10. Dairies	.310174	.326446	.016272	5.25
11. Bakeries	.318405	.327016	.008611	2.70
12. Beverages	.644361	.662528	.018167	2.87
17. Chemicals	.749271	.798360	.049098	6.55
18. Petroleum products	.804049	.819796	.015747	1.96
19. Glass	.418595	.450058	.031463	7.52
20. Stone & clay products	.784269	.810842	.026573	3.39
21. Primary metals	.688787	.835293	.146506	21.27
22. Fabricated metals	.307165	.338806	.031641	10.30
23. Machinery (except electric)	.229605	.252761	.023156	10.08
24. Electric machinery	.328578	.339549	.010971	3.34
25. Transportation equipment	.516487	.545057	.028570	5.53
26. Instruments & products	.245097	.275846	.029749	12.09
46. Electrical systems	1.751283	2.039464	.288161	16.45%

computational purposes the relevant coefficients are the elements of the capital expansion (or B) matrix that are given in Appendix Table 3A-1.

As one might have inferred from Table 3-1, the largest absolute increase in an individual sector's total capital coefficient is found in the electric utility sector. The generation of electric energy is a highly capital-intensive activity, and the expansion capital coefficients of this sector are quite large compared with most other sectors.[9] Because of this high degree of capital intensity, the *relative* impact on the capital coefficients of the electric utility sector is smaller than that on the primary metals coefficients. The total expansion capital coefficient of primary metals will increase by more than 21 percent compared with an increase of about 16 percent for electric utilities. Despite the sizeable investment in pollution abatement equipment that will be required in the chemicals sector, the total expansion capital coefficient here will go up by less than 7 percent. The absolute and relative changes for other sectors are given in the right hand columns of Table 3-2.

The increases in total replacement capital coefficients due to particulate abatement investment are presented in Table 3-3. The changes in this table roughly parallel those of Table 3-2 and require little further comment. It may be worth pointing out, however, that the expansion capital coefficients depend upon *capacity* in the individual sectors while the replacement coefficients are functions of the level of output in those sectors. Thus, it is possible for expansion capital coefficients to be greater than one (as are the expansion capital coefficients of electric utilities), but replacement capital coefficients are

Table 3-3
Changes in Total Replacement Capital Coefficients (D_j) Due to Particulate Abatement Investment

	Original D_{ij}	New D_{ij}	Difference	Per Cent Change
2. Coal mines (undgd.) 3. Coal (strip & auger)	.033860	.034106	.000246	.73%
9. Food products (n.e.c.)	.018490	.019028	.000538	2.91
10. Dairies	.015210	.015580	.000370	2.43
11. Bakeries	.016980	.017439	.000459	2.70
12. Beverages	.048410	.049769	.001359	2.81
17. Chemicals	.036380	.039058	.002678	7.36
18. Petroleum products	.042680	.043515	.000835	1.95
19. Glass	.018740	.020129	.001389	7.41
20. Stone & clay products	.048300	.049855	.001555	3.22
21. Primary metals	.038410	.046577	.008167	21.26
22. Fabricated metals	.023580	.026004	.002424	10.27
23. Machinery (except electric)	.010520	.011672	.001152	10.95
24. Electric machinery	.017705	.018714	.001009	5.70
25. Transportation equipment	.061298	.066207	.004909	8.01
26. Instruments & products	.023370	.025944	.002594	11.10
46. Electrical systems	.053520	.061376	.007856	14.68%

all less than unity. The management of an individual establishment will be willing to spend more than one dollar to expand his *capacity* by one dollar's worth since this increased capacity will add to output throughout its useful life. But capital will not be replaced as it is worn out unless there is sufficient demand for the firm's product or service to warrant the replacement of equipment as it is "used up" in the production process. As in the case of the expansion capital coefficients, the detailed changes in the elements of the replacement capital coefficient matrix are given in Appendix Table 3A-2.

Changes in *total* interindustry (technical or input) coefficients as a result of the installation and operation of particulate abatement equipment are quite small compared with the changes in capital coefficients discussed above. These changes are summarized in Table 3-4. Of the major polluters, the chemicals sector is the only one to register an increase of more than one percent. In primary metals, the increase is negligible and in electric utilities, the total interindustry coefficient goes up about 0.7 percent. In a few sectors the total interindustry coefficients register small negative changes. These declines are offset, however, by increases in imports or other primary purchases.

The import coefficients of West Virginia industry will be affected to some extent by the installation and operation of particulate abatement equipment (see Table 3-5). As in the case of interindustry coefficients, most of the changes are small. The only increase of any size is in the electric utilities coefficient, which is expected to go up by almost 5 percent.

The effects of particulate abatement on total household coefficients are given

Table 3-4
Changes in Total Interindustry Coefficients (A_j) Due to Particulate Abatement

		Original A_{ij}	New A_{ij}	Difference	Per Cent Change
2.	Coal mines (undgd.)	.114069	.114615	.000006	ϕ%
9.	Food products (n.e.c.)	.444880	.447979	.003099	.70
10.	Dairies	.392360	.392295	−.000065	−.02
11.	Bakeries	.227350	.227310	.000040	−.02
12.	Beverages	.160830	.160965	.000135	.08
17.	Chemicals	.278130	.281240	.003110	1.12
18.	Petroleum products	.348590	.350945	.002355	.68
19.	Glass	.258222	.268119	.009897	.09
20.	Stone & clay products	.344030	.343799	−.000231	−.07
21.	Primary metals	.097739	.098781	.001042	1.07
22.	Fabricated metals	.181980	.182149	.000169	.09
23.	Machinery (except electric)	.147800	.148000	.000200	.14
24.	Electric machinery	.318270	.318213	−.000057	−.02
25.	Transportation equipment	.131250	.135117	.003867	2.95
26.	Instruments & products	.101926	.105530	.003604	3.53
46.	Electrical systems	.256380	.258265	.001885	.74%

ϕLess than .01 of 1%

Table 3-5
Changes in Import Coefficients (a_{mj}) Due to Particulate Abatement

	Original M_{ij}	New M_{ij}	Difference	Per Cent Change
2. Coal mines (undgd.)	.338430	.338424	-.000006	ϕ %
9. Food products (n.e.c.)	.384150	.387146	.002996	.78
10. Dairies	.374830	.374933	.000103	.03
11. Bakeries	.430870	.431025	.000155	.04
12. Beverages	.326810	.327057	.000247	.08
17. Chemicals	.372450	.374719	.002269	.61
18. Petroleum products	.376310	.378270	.001960	.52
19. Glass	.179980	.180954	.000954	.54
20. Stone & clay products	.175330	.176482	.001152	.66
21. Primary metals	.567890	.568077	.000187	.03
22. Fabricated metals	.433560	.434064	.000504	.12
23. Machinery (except electric)	.441750	.442056	.000306	.07
24. Electric machinery	.267420	.267693	.000213	.08
25. Transportation equipment	.489880	.487953	-.001927	-.39
26. Instruments & products	.522310	.520382	-.001923	-.37
46. Electrical systems	.152810	.160010	.007200	4.71 %

ϕLess than .01 of 1%

in Table 3-6. In all but a few sectors household coefficients–which primarily represent labor inputs–register declines. This does not mean that less labor will be required by establishments after they install particulate abatement equipment. Indeed, in many cases, additional labor will have to be employed to operate and maintain that equipment. What Table 3-6 shows is the *relative* declines in labor inputs associated with relative increases in capital intensity. This means that increases in other inputs–both interindustry and primary–are greater than the increases in labor inputs associated with particulate pollution abatement efforts.

The same is true of the catch-all "all other" final payments coefficients summarized in Table 3-7. The declines in most of these sectoral coefficients indicate that final payments in this category–such as state, federal, and local taxes–will not increase as rapidly as interindustry transactions or imports, if they increase at all.

The total coefficients summarized in Table 3-4 through 3-7 are useful only because they present a bird's eye view of changes in input coefficients on current account. The detailed changes on a sector-by-sector basis are given in Appendix Table 3A-3.

The Direct and Indirect Cost of Particulate Abatement

This section presents the final step in the analysis of the cost of particulate abatement. We saw earlier (in Table 3-1) that the total capital cost of installing

Table 3-6
Changes in Household Coefficients (a_{hj}) Due to Particulate Abatement

	Original H_{ij}	New H_{ij}	Difference	Per Cent Change
2. Coal mines (undgd.)	.401910	.401604	-.000306	-.08%
9. Food products (n.e.c.)	.128180	.129152	.000972	.76
10. Dairies	.183370	.183370	0	0
11. Bakeries	.296230	.296235	.000005	φ
12. Beverages	.257750	.257660	-.000090	.04
17. Chemicals	.156220	.156080	-.000140	-.09
18. Petroleum products	.110450	.110002	-.000448	-.41
19. Glass	.303420	.303000	-.000402	-.14
20. Stone & clay products	.294200	.293822	-.000378	-.13
21. Primary metals	.234490	.233182	-.001308	-.56
22. Fabricated metals	.289600	.289171	-.000429	-.15
23. Machinery (except electric)	.274430	.274213	-.000217	-.08
24. Electric machinery	.271640	.271536	-.000104	-.04
25. Transportation equipment	.255040	.253800	-.001240	-.41
26. Instruments & products	.289590	.287005	-.002585	-.89
46. Electrical systems	.206190	.203567	-.002623	-1.27%

φLess than .01 of 1%

particulate abatement equipment, in 1975 dollars, was $309.4 million and that the annual operating costs, again in 1975 dollars, would come to $37.8 million. Thus, the total direct expenditures on pollution abatement equipment, plus the operating costs for one year, would be $347.2 million.[10]

The changes in output in each sector that would be required to produce a constant bill of deliveries to final demand, after the installation of particulate abatement equipment, are given in the third column of Table 3-8. These are the measures we have used to estimate the "real costs" of meeting clean air standards. They represent the value of additional resources that would be required to maintain a constant level of output. In absolute terms, there are fairly large changes in the chemicals, primary metals, and electric utility sectors—as expected in the case of these heavy polluters. But there are large changes for general and special trades contractors, which reflect the increases in capital spending by other sectors. There are also large increases in stone and clay products and in fabricated metal products, once again due to the effects of capital spending elsewhere on particulate abatement plant and equipment. For the same reason, the banking sector shows a substantial increase. Changes in most other sectors are small to moderate, and in a few sectors small declines are registered. Negative changes are the result of input substitution, but in no case is the decline large enough to be of economic significance.

Interindustry transactions would increase by $116.4 million, or 2.3 percent. Imports would also go up by $146.3 million, or 2.4 percent. When all the direct and indirect effects on both interindustry transactions and primary payments are taken into account, the cost of particulate abatement, in 1975 dollars, would be $271.5 million. The West Virginia economy would have to generate an increase

Table 3-7
Changes in "All Other" Final Payments Coefficients (a_{fi}) Due to Particulate Abatement

	Original F_{ij}	New F_{ij}	Difference	Per Cent Change
2. Coal Mines (undgd.)	.145000	.144901	-.000099	-.07%
9. Food products (n.e.c.)	.042790	.043153	.000363	.85
10. Dairies	.049440	.049452	.000012	.02
11. Bakeries	.045550	.045526	-.000024	-.05
12. Beverages	.254610	.254445	-.000165	-.06
17. Chemicals	.193200	.193326	.000126	.06
18. Petroleum products	.164650	.165215	.000565	.34
19. Glass	.258370	.257876	.000494	-.19
20. Stone & clay products	.186440	.186112	-.000328	-.18
21. Primary metals	.099880	.099239	-.000064	-.64
22. Fabricated metals	.094940	.094891	-.000049	-.06
23. Machinery (except electric)	.136020	.135932	-.000088	-.06
24. Electric machinery	.142670	.142624	-.000046	-.03
25. Transportation equipment	.123830	.123240	-.000590	-.48
26. Instruments & products	.086270	.086224	-.000046	-.05
46. Electrical systems	.384620	.378156	-.006464	-1.68

of 1.34 percent in total output to produce a constant bill of goods for delivery to final demand and, at the same time, meet the clean air requirements established by West Virginia and federal laws.

Table 3-8
Direct and Indirect Cost of Particulate Abatement (Thousands of 1975 Dollars)

	Original TGO minus Final Demand	New TGO minus Final Demand	Changes	
			Dollars	Per Cent
1. Agriculture	$ 80,879	$ 81,002	$ 123	.15%
2. Coal mines (undgd.)	176,735	179,529	2,794	1.58
3. Coal (strip & auger)	71,275	71,009	-226	-.37
4. Petroleum & natural gas	130,461	130,657	196	.15
5. All other mining	84,747	85,483	736	.87
6. Building contractors	508,428	553,170	44,742	8.80
7. Non-building contractors	165,499	169,209	3,710	2.24
8. Special contractors	337,323	354,708	17,385	5.15
9. Food products (n.e.c.)	33,214	33,289	75	.22
10. Dairies	16,157	16,167	10	.06
11. Bakeries	9,805	9,812	7	.07
12. Beverages	5,707	5,710	3	.05
13. Apparel & accessories	6,082	6,089	7	.12
14. Logging & sawmills	60,816	61,695	879	1.44
15. Furn. & wood fabrication	25,534	26,367	833	3.26
16. Printing & publishing	163,837	164,374	537	.33
17. Chemicals	427,068	431,936	4,868	1.14
18. Petroleum products	99,089	99,785	696	.70
19. Glass	28,435	28,439	4	.01
20. Stone & clay products	102,766	107,706	494	4.81
21. Primary metals	179,736	183,592	3,856	2.14
22. Fabricated metals	103,023	106,947	3,924	3.81
23. Machinery (except electric)	138,843	145,465	6,622	4.77
24. Electric machinery	35,244	36,037	793	2.25
25. Transportation equipment	6,223	6,258	35	.56
26. Instruments & products	8,517	8,513	-4	-.05
27. All other manufacturing	$ 106,525	$ 107,331	$ 806	.76%

Table 3-8 (cont.)

	Original TGO minus Final Demand	New TGO minus Final Demand	Changes	
			Dollars	Per Cent
28. Restaurants & bars	$ 13,631	$ 13,800	$ 169	1.24%
29. Wholesale trade	230,899	232,702	1,803	.78
30. Retail food stores	2,032	2,034	2	.10
31. Auto service stations	28,782	28,958	176	.61
32. All other retail	125,249	126,618	1,369	1.09
33. Banking	71,641	79,447	7,806	10.90
34. Other finance	15,800	15,875	75	.47
35. Insurance agents	163,922	164,819	897	.55
36. Real estate	63,782	64,288	506	.79
37. All other FIRE	65,723	65,941	218	.33
38. Hotels & lodgings	14,382	14,418	36	.25
39. Medical & legal services	105,899	106,239	340	.32
40. Educational services	8,796	8,802	6	.07
41. All other services	262,326	263,230	904	.34
42. Railroads	80,775	80,671	-104	-.13
43. Trucking & warehousing	164,433	165,285	852	.52
44. All other transportation	76,757	76,648	-109	-.14
45. Communications	121,848	122,126	278	.23
46. Electrical systems	207,715	212,087	4,372	2.10
47. Gas systems	159,513	162,089	1,576	.99
48. Water & sanitary services	38,891	40,240	1,349	3.47
Total interindustry transactions	5,140,170	5,256,596	116,426	2.26
Households	6,017,364	6,023,765	6,401	.11
Out-of-state (imports)	6,079,203	6,225,455	146,252	2.41
All other final payments	3,059,643	3,062,060	2,417	.08
Total final payments	15,156,210	15,311,280	155,070	1.02
Total change in TGO	$20,296,380	$ 20,567,876	$271,496	1.34%

Appendix 3A

Table 3A-1
Changes in Expansion Capital Coefficients Due to Particulate Abatement Investment

	Coal Mining (2 & 3)			Food & Kindred Products, n.e.c. (9)		
	Original b_{ij}	New b_{ij}	Difference	Original b_{ij}	New b_{ij}	Difference
6. Building contractors	.140950	.142008	.001058	.091834	.094811	.002977
7. Non-building contractors	.006473	.006522	.000049	.005755	.005942	.000187
8. Special contractors	.001595	.001607	.000012	.005672	.005856	.000184
14. Logging & sawmills	.002503	.002522	.000019	.011409	.011779	.000370
15. Furn. & wood fabrication						
20. Stone & clay products	.013662	.013765	.000103	.011428	.011798	.000370
21. Primary metals	.003732	.003760	.000028	.011541	.011915	.000374
22. Fabricated metals				.005658	.005841	.000183
23. Machinery (except electric)	.069088	.069607	.000519			
24. Electric machinery				.006195	.006396	.000201
25. Transportation equipment	.001483	.001494	.000011			
26. Instruments & products						
27. All other manufacturing						
29. Wholesale trade	.002274	.002291	.000017	.005617	.005799	.000182
32. All other retail	.001776	.001789	.000013	.017258	.017817	.000559
49. Out-of-state (imports)	.443032	.446357	.003325	.145925	.150655	.004730

Table 3A-1 (cont.)

	Dairies (10)			Bakeries (11)		
	Original b_{ij}	New b_{ij}	Difference	Original b_{ij}	New b_{ij}	Difference
6. Building contractors	.151215	.154936	.003721	.130416	.133945	.003529
7. Non-building contractors	.009477	.009707	.000230	.002179	.002238	.000059
8. Special contractors	.009340	.009568	.000228	.021480	.022061	.000581
14. Logging & sawmills	.009394	.009624	.000230	.010802	.011088	.000286
15. Furn. & wood fabrication						
20. Stone & clay products	.009409	.009639	.000230	.010820	.011113	.000293
21. Primary metals						
22. Fabricated metals				.010713	.011003	.000290
23. Machinery (except electric)						
24. Electric machinery	.005750	.005891	.000141	.003199	.003286	.000087
25. Transportation equipment						
26. Instruments & products						
27. All other manufacturing						
29. Wholesale trade	.018496	.018948	.000452	.010635	.010923	.000288
32. All other retail	.009287	.009514	.000227	.008970	.009213	.000243
49. Out-of-state (imports)	.096265	.098619	.002354	.109191	.112146	.002955

Table 3A-1 (cont.)

	Beverages (12)			Chemicals (17)		
	Original b_{ij}	New b_{ij}	Difference	Original b_{ij}	New b_{ij}	Difference
6. Building contractors	.047493	.048832	.001339	.267804	.289334	.021530
7. Non-building contractors				.138557	.138557	0
8. Special contractors	.031712	.032606	.000894			
14. Logging & sawmills	.031894	.032793	.000899			
15. Furn. & wood fabrication						
20. Stone & clay products	.031946	.032847	.000901	.012753	.013778	.001025
21. Primary metals				.025286	.027319	.002033
22. Fabricated metals	.031630	.032522	.000892			
23. Machinery (except electric)				.103712	.112049	.008337
24. Electric machinery				.016700	.018043	.001343
25. Transportation equipment				.000404	.000437	.000033
26. Instruments & products						
27. All other manufacturing						
29. Wholesale trade	.031401	.032286	.000885	.001540	.001664	.000124
32. All other retail	.058018	.059654	.001636	.006493	.007015	.000522
49. Out-of-state (imports)	.380267	.390988	.010721	.176022	.190173	.014151

Table 3A-1 (cont.)

	Petroleum (18)			Glass (19)		
	Original b_{ij}	New b_{ij}	Difference	Original b_{ij}	New b_{ij}	Difference
6. Building contractors	.181427	.184980	.003553	.166424	.178765	.012341
7. Non-building contractors				.061220	.065760	.004540
8. Special contractors	.044822	.045700	.000878	.004830	.005188	.000358
14. Logging & sawmills						
15. Furn. & wood fabrication				.004096	.004400	.000304
20. Stone & clay products						
21. Primary metals	.006384	.006509	.000125			
22. Fabricated metals	.041131	.041937	.000806	.002725	.002927	.000202
23. Machinery (except electric)				.021996	.023627	.001631
24. Electric machinery				.002909	.003125	.000216
25. Transportation equipment	.004480	.004568	.000088			
26. Instruments & products						
27. All other manufacturing						
29. Wholesale trade	.009764	.009955	.000191	.003921	.004212	.000291
32. All other retail	.092701	.094516	.001815	.002244	.002410	.000166
49. Out-of-state (imports)	.423340	.431631	.008291	.148623	.159644	.011021

Table 3A-1 (cont.)

	Stone & Clay (20)			Primary Metals (21)		
	Original b_{ij}	New b_{ij}	Difference	Original b_{ij}	New b_{ij}	Difference
6. Building contractors	.085205	.088092	.002887	.132210	.160214	.028004
7. Non-building contractors	.004070	.004208	.000138	.051069	.061921	.010852
8. Special contractors	.004011	.004147	.000136	.057900	.070203	.012303
14. Logging & sawmills						
15. Furn. & wood fabrication				.001987	.002409	.000422
20. Stone & clay products				.032448	.039556	.007108
21. Primary metals				.036235	.043935	.007700
22. Fabricated metals	.000200	.000207	.000007	.026049	.031584	.005535
23. Machinery (except electric)	.066049	.068287	.002238	.081213	.098470	.017257
24. Electric machinery						
25. Transportation equipment						
26. Instruments & products						
27. All other manufacturing	.008283	.008564	.000281	.004519	.005479	.000960
29. Wholesale trade	.011162	.011540	.000378	.007837	.009502	.001665
32. All other retail	.085039	.087920	.002881	.000866	.001050	.000184
49. Out-of-state (imports)	.520250	.537877	.017627	.256454	.310970	.054516

Table 3A-1 (cont.)

	Fabricated Metals (22)			Machinery (non-electrical) (23)		
	Original b_{ij}	New b_{ij}	Difference	Original b_{ij}	New b_{ij}	Difference
6. Building contractors	.025679	.027133	.001454	.096842	.107466	.010624
7. Non-building contractors	.004292	.004738	.000446	.008736	.009697	.000961
8. Special contractors	.004230	.004670	.000440	.004305	.004777	.000472
14. Logging & sawmills	.004254	.004696	.000442			
15. Furn. & wood fabrication	.002123	.002344	.000221	.001081	.001192	.000111
20. Stone & clay products	.004261	.004704	.000443			
21. Primary metals				.035389	.039272	.003883
22. Fabricated metals				.000129	.000143	.000014
23. Machinery (except electric)	.000685	.000756	.000071	.015110	.016768	.001658
24. Electric machinery	.000419	.000463	.000044	.014191	.015748	.001557
25. Transportation equipment	.000676	.000746	.000070			
26. Instruments & products						
27. All other manufacturing						
29. Wholesale trade	.008376	.009247	.000871	.004263	.004731	.000468
32. All other retail	.001771	.001955	.000184	.003256	.003613	.000357
49. Out-of-state (imports)	.250959	.277054	.026095	.046277	.051354	.005077

Table 3A-1 (cont.)

	Electrical Machinery (24)			Transportation Equipment (25)		
	Original b_{ij}	New b_{ij}	Difference	Original b_{ij}	New b_{ij}	Difference
6. Building contractors	.151872	.156942	.005070	.117656	.124020	.006364
7. Non-building contractors				.014746	.015544	.000798
8. Special contractors				.014533	.015319	.000786
14. Logging & sawmills						
15. Furn. & wood fabrication	.002107	.002177	.000070	.014596	.015386	.000790
20. Stone & clay products				.014641	.015433	.000792
21. Primary metals				.014786	.015586	.000800
22. Fabricated metals				.014496	.015280	.000784
23. Machinery (except electric)	.003655	.003777	.000122	.029416	.031007	.001591
24. Electric machinery	.001666	.001722	.000056	.010101	.010647	.000546
25. Transportation equipment	.004360	.004506	.000146			
26. Instruments & products						
27. All other manufacturing						
29. Wholesale trade	.005408	.005590	.000182	.028783	.030340	.001557
32. All other retail	.014865	.015361	.000496	.002890	.003046	.000156
49. Out-of-state (imports)	.144645	.149474	.004829	.240443	.253449	.013006

Table 3A-1 (cont.)

		Instruments & Related Products (26)			Electric Utilities (46)		
		Original b_{ij}	New b_{ij}	Difference	Original b_{ij}	New b_{ij}	Difference
6.	Building contractors	.062357	.069227	.006870	.472151	.627537	.155386
7.	Non-building contractors	.006252	.006941	.000689	.059093	.059093	0
8.	Special contractors	.006162	.006845	.000683	.044021	.044021	0
14.	Logging & sawmills						
15.	Furn. & wood fabrication	.006189	.006871	.000682			
20.	Stone & clay products	.006208	.006892	.000684	.013883	.018775	.004892
21.	Primary metals				.015275	.024972	.009697
22.	Fabricated metals						
23.	Machinery (except electric)				.045257	.045257	0
24.	Electric machinery)				.003631	.003631	0
25.	Transportation equipment						
26.	Instruments & products				.001901	.001901	0
27.	All other manufacturing						
29.	Wholesale trade	.012204	.013549	.001345	.020623	.021227	.000604
32.	All other retail	.006127	.006802	.000675	.000438	.002941	.002503
49.	Out-of-state (imports)	.142968	.158719	.015751	1.075010	1.190109	.115099

Table 3A-2
Changes in Replacement Capital Coefficients Due to Investment in Particulate Abatement

		Coal Mining (2 & 3)			Food & Kindred Products, n.e.c. (9)		
		Original d_{ij}	New d_{ij}	Difference	Original d_{ij}	New d_{ij}	Difference
5.	All other mining	.003130	.003153	.000023	.001530	.001580	.000050
6.	Building contractors	.000140	.000141	.000001	.000100	.000103	.000003
7.	Non-building contractors	.000159	.000161	.000002	.000470	.000485	.000015
8.	Special contractors						
13.	Apparel & accessories	.000250	.000252	.000002	.000950	.000981	.000031
14.	Logging & sawmills						
15.	Furn. & wood fabrication						
20.	Stone & clay products	.000300	.000302	.000002	.000190	.000196	.000006
21.	Primary metals	.000080	.000081	.000001	.000190	.000196	.000006
22.	Fabricated metals				.000470	.000485	.000015
23.	Machinery (except electric)	.006909	.006961	.000052			
24.	Electric machinery				.000520	.000536	.000016
25.	Transportation equipment	.000148	.000149	.000001			
26.	Instruments & products						
27.	All other manufacturing						
29.	Wholesale trade	.000227	.000229	.000002	.000470	.000485	.000015
32.	All other retail	.000178	.000179	.000001	.001440	.001487	.000047
49.	Out-of-state (imports)	.022330	.022498	.000168	.012160	.012554	.000394

Table 3A-2 (cont.)

	Food & Kindred Products, dairies (10)			Food & Kindred Products, bakeries (11)		
	Original d_{ij}	New d_{ij}	Difference	Original d_{ij}	New d_{ij}	Difference
5. All other mining	.002520	.002582	.000062			
6. Building contractors	.000160	.000164	.000004	.002170	.002229	.000059
7. Non-building contractors	.000780	.000799	.000019	.000040	.000041	.000001
8. Special contractors				.001790	.001838	.000048
13. Apparel & accessories	.000780	.000799	.000019			
14. Logging & sawmills				.000900	.000924	.000024
15. Furn. & wood fabrication						
20. Stone & clay products	.000160	.000164	.000004	.000180	.000185	.000005
21. Primary metals						
22. Fabricated metals				.000890	.000914	.000024
23. Machinery (except electric)						
24. Electric machinery	.000480	.000492	.000012	.000270	.000277	.000007
25. Transportation equipment						
26. Instruments & products						
27. All other manufacturing						
29. Wholesale trade	.001540	.001577	.000037	.000890	.000915	.000025
32. All other retail	.000770	.000789	.000019	.000750	.000770	.000020
49. Out-of-state (imports)	.008020	.008216	.000196	.009100	.009346	.000246

Table 3A-2 (cont.)

	Food & Kindred Products, beverages (12)			Chemicals (17)		
	Original d_{ij}	New d_{ij}	Difference	Original d_{ij}	New d_{ij}	Difference
5. All other mining						
6. Building contractors	.000790	.000812	.000022	.005950	.006428	.000478
7. Non-building contractors				.003080	.003080	0
8. Special contractors	.002640	.002715	.000075			
13. Apparel & accessories						
14. Logging & sawmills	.002660	.002735	.000075			
15. Furn. & wood fabrication						
20. Stone & clay products	.000530	.000545	.000015	.000280	.000303	.000023
21. Primary metals				.000560	.000605	.000045
22. Fabricated metals	.002640	.002714	.000074			
23. Machinery (except electric)				.009430	.010188	.000758
24. Electric machinery				.001520	.001642	.000122
25. Transportation equipment				.000040	.000044	.000004
26. Instruments & products						
27. All other manufacturing						
29. Wholesale trade	.002620	.002694	.000074	.000140	.000151	.000011
32. All other retail	.004835	.004971	.000136	.000590	.000638	.000048
49. Out-of-state (imports)	.031690	.032583	.000893	.014790	.015979	.001189

Table 3A-2 (cont.)

	Petroleum (18)			Glass (19)		
	Original d_{ij}	New d_{ij}	Difference	Original d_{ij}	New d_{ij}	Difference
5. All other mining						
6. Building contractors	.004030	.004109	.000079	.003400	.003974	.000274
7. Non-building contractors				.001360	.001461	.000101
8. Special contractors	.002800	.002855	.000055	.000350	.000376	.000076
13. Apparel & accessories						
14. Logging & sawmills						
15. Furn. & wood fabrication				.000290	.000311	.000021
20. Stone & clay products						
21. Primary metals	.000140	.000143	.000003			
22. Fabricated metals	.002570	.002620	.000050	.000200	.000215	.000015
23. Machinery (except electric)				.001570	.001686	.000116
24. Electric machinery				.000210	.000225	.000015
25. Transportation equipment	.000280	.000285	.000005			
26. Instruments & products						
27. All other manufacturing						
29. Wholesale trade	.000610	.000622	.000012	.000280	.000301	.000021
32. All other retail	.005790	.005903	.000113	.000160	.000172	.000012
49. Out-of-state (imports)	.026460	.026978	.000518	.010620	.011408	.000788

Table 3A-2 (cont.)

	Stone & Clay Products (20)			Primary Metal Products (21)		
	Original d_{ij}	New d_{ij}	Difference	Original d_{ij}	New d_{ij}	Difference
5. All other mining						
6. Building contractors	.001890	.001954	.000064	.002940	.003565	.000625
7. Non-building contractors	.000090	.000012	.000003	.001130	.001370	.000240
8. Special contractors	.000270	.000279	.000009	.003620	.004389	.000769
13. Apparel & accessories						
14. Logging & sawmills						
15. Furn. & wood fabrication				.000120	.000145	.000025
20. Stone & clay products				.000720	.000878	.000158
21. Primary metals				.000810	.000982	.000172
22. Fabricated metals	.000010	.000010	0	.001630	.001976	.000346
23. Machinery (except electric)	.004400	.004549	.000149	.005080	.006160	.001080
24. Electric machinery						
25. Transportation equipment						
26. Instruments & products						
27. All other manufacturing	.000550	.000569	.000019	.000280	.000339	.000059
29. Wholesale trade	.000740	.000765	.000025	.000490	.000594	.000104
32. All other retail	.005670	.005862	.000192	.000050	.000060	.000010
49. Out-of-state (imports)	.034680	.035855	.001175	.021540	.026119	.004579

Table 3A-2 (cont.)

	Fabricated Metal Products (22)			Machinery (non-electrical) (23)		
	Original d_{ij}	New d_{ij}	Difference	Original d_{ij}	New d_{ij}	Difference
5. All other mining						
6. Building contractors	.000570	.000602	.000032	.002150	.002386	.000236
7. Non-building contractors	.000100	.000110	.000010	.000190	.000211	.000021
8. Special contractors	.000350	.000386	.000036	.000359	.000398	.000039
13. Apparel & accessories						
14. Logging & sawmills	.000360	.000398	.000038			
15. Furn. & wood fabrication	.000180	.000199	.000019	.000090	.000099	.000009
20. Stone & clay products	.000100	.000110	.000010			
21. Primary metals				.000790	.000877	.000087
22. Fabricated metals				.000010	.000011	.000001
23. Machinery (except electric)	.000060	.000067	.000007	.001260	.001398	.000138
24. Electric machinery	.000040	.000044	.000004	.001180	.001310	.000130
25. Transportation equipment	.000060	.000066	.000006			
26. Instruments & products						
27. All other manufacturing						
29. Wholesale trade	.000700	.000772	.000072	.000360	.000400	.000040
32. All other retail	.000150	.000166	.000016	.000270	.000299	.000029
49. Out-of-state (imports)	.020910	.023084	.002174	.003860	.004283	.000423

Table 3A-2 (cont.)

		Electrical Machinery (24)			Transportation Equipment (25)		
		Original d_{ij}	New d_{ij}	Difference	Original d_{ij}	New d_{ij}	Difference
5.	All other mining						
6.	Building contractors	.003380	.003493	.000113	.002620	.002762	.000142
7.	Non-building contractors				.000330	.000348	.000018
8.	Special contractors				.002420	.002551	.000131
13.	Apparel & accessories						
14.	Logging & sawmills						
15.	Furn. & wood fabrication	.000180	.000186	.000006	.002429	.002561	.000132
20.	Stone & clay products				.000330	.000348	.000018
21.	Primary metals				.000330	.000348	.000018
22.	Fabricated metals				.002419	.002550	.000131
23.	Machinery (except electric)	.000310	.000320	.000010	.004899	.005164	.000265
24.	Electric machinery	.000140	.000144	.000004	.001680	.001771	.000091
25.	Transportation equipment	.000360	.000372	.000012			
26.	Instruments & products						
27.	All other manufacturing						
29.	Wholesale trade	.000450	.000465	.000015	.004800	.005060	.000260
32.	All other retail	.001240	.001281	.000041	.000480	.000506	.000026
49.	Out-of-state (imports)	.012050	.012453	.000403	.040071	.042238	.002167

Table 3A-2 (cont.)

		Instruments & Related Products (26)			Electric Companies & Systems (46)		
		Original d_{ij}	New d_{ij}	Difference	Original d_{ij}	New d_{ij}	Difference
5.	All other mining				.010490	.013942	.003452
6.	Building contractors	.001390	.001543	.000153	.001310	.001310	0
7.	Non-building contractors	.000140	.000155	.000015	.001520	.001520	0
8.	Special contractors	.000770	.000856	.000086			
13.	Apparel & accessories						
14.	Logging & sawmills						
15.	Furn. & wood fabrication	.000770	.000855	.000085			
20.	Stone & clay products	.000140	.000155	.000015	.000310	.000419	.000109
21.	Primary metals				.000339	.000554	.000215
22.	Fabricated metals						
23.	Machinery (except electric)				.001560	.001560	0
24.	Electric machinery				.000130	.000130	0
25.	Transportation equipment						
26.	Instruments & products				.000070	.000070	0
27.	All other manufacturing						
29.	Wholesale trade	.001521	.001688	.000167	.000710	.000731	.000021
32.	All other retail	.000770	.000854	.000084	.000015	.000101	.000086
49.	Out-of-state (imports)	.017870	.019838	.001968	.037070	.041039	.003969

Table 3A-3
Changes in Input Coefficients Due to Particulate Abatement Investment

	Coal Mines (undgd.) (2)		
	Original a_{ij}	New a_{ij}	Difference
1. Agriculture			
2. Coal Mines (undgd.)	.009545	.009553	.000008
3. Coal (strip & auger)	.008926	.008929	.000003
4. Petroleum & natural gas			
5. All other mining	.000838	.000838	0
6. Building contractors	0	.000003	.000003
7. Non-building contractors			
8. Special contractors	.004417	.004422	.000005
9. Food products (n.e.c.)	.001041	.001042	.000001
10. Dairies			
11. Bakeries			
12. Beverages			
13. Apparel & accessories			
14. Logging & sawmills	.006483	.006486	.000003
15. Furn. & wood fabrication	.003655	.003657	.000002
16. Printing & publishing	.000786	.000787	.000001
17. Chemicals	.000338	.000377	.000039
18. Petroleum products	.001837	.001838	.000001
19. Glass			
20. Stone & clay products	.003807	.003808	.000001
21. Primary metals	.000107	.000111	.000004
22. Fabricated metals	.000790	.000791	.000001
23. Machinery (except electric)	.000556	.000556	0
24. Electric machinery			
25. Transportation equipment			
26. Instruments & products			
27. All other manufacturing	.000981	.000098	-.000883
28. Restaurants & bars			
29. Wholesale trade	.009448	.009458	.000010
30. Retail food stores			
31. Auto service stations	.000020	.000020	0
32. All other retail	.002861	.002862	.000001
33. Banking	.000037	.000076	.000039
34. Other finance			
35. Insurance agents	.015388	.015394	.000006
36. Real estate			
37. All other FIRE	.002369	.002370	.000001
38. Hotels & lodgings			
39. Medical & legal services	.000576	.000576	0
40. Educational services			
41. All other services	.003436	.003437	.000001
42. Railroads	.000240	.000243	.000003
43. Trucking & warehousing	.001149	.001155	.000006
44. All other transportation	.000050	.000050	0
45. Communications	.001015	.001015	0
46. Electrical systems	.033874	.033896	.000022
47. Gas systems	.000039	.000039	0
48. Water & sanitary services			
49. Households	.401910	.401604	-.000306
50. Out-of-state (imports)	.338430	.338424	-.000006
51. All other final payments	.145000	.144901	-.000099
52. Total	1.000000	.999544	-.000456

Table 3A-3 (cont.)

	Food products (n.e.c.) (9)		
	Original a_{ij}	New a_{ij}	Difference
1. Agriculture	.142935	.143895	.000960
2. Coal Mines (undgd.)	.000015	.000015	0
3. Coal (strip & auger)			
4. Petroleum & natural gas			
5. All other mining	.000146	.000147	.000001
6. Building contractors			
7. Non-building contractors			
8. Special contractors	.000849	.000855	.000006
9. Food products (n.e.c.)	.075085	.075589	.000504
10. Dairies			
11. Bakeries			
12. Beverages			
13. Apparel & accessories			
14. Logging & sawmills			
15. Furn. & wood fabrication			
16. Printing & publishing	.000659	.000663	.000004
17. Chemicals	0	.000059	.000059
18. Petroleum products			
19. Glass			
20. Stone & clay products	.000044	.000044	0
21. Primary metals			
22. Fabricated metals			
23. Machinery (except electric)			
24. Electric machinery			
25. Transportation equipment			
26. Instruments & products	.010427	.010497	.000070
27. All other manufacturing	.073093	.066213	-.006880
28. Restaurants & bars	.000146	.000147	.000001
29. Wholesale trade	.017442	.017574	.000132
30. Retail food stores			
31. Auto service stations	.001596	.001607	.000011
32. All other retail	.004291	.004320	.000029
33. Banking	.002065	.002138	.000073
34. Other finance	.020210	.020346	.000136
35. Insurance agents	.001992	.002005	.000013
36. Real estate			
37. All other FIRE			
38. Hotels & lodgings	.000205	.000206	.000001
39. Medical & legal services	.006532	.006576	.000044
40. Educational services	.000644	.000649	.000005
41. All other services	.019112	.019240	.000128
42. Railroads			
43. Trucking & warehousing	.001128	.001150	.000022
44. All other transportation			
45. Communications	.040112	.040382	.000270
46. Electrical systems	.023403	.023575	.000172
47. Gas systems	.000996	.001003	.000007
48. Water & sanitary services	.001699	.001710	.000011
49. Households	.128180	.129152	.000972
50. Out-of-state (imports)	.384150	.387146	.002996
51. All other final payments	.042790	.043153	.000363
52. Total	1.000000	1.007430	.007430

Table 3A-3 (cont.)

	Dairies (10)		
	Original a_{ij}	New a_{ij}	Difference
1. Agriculture	.294805	.294689	-.000116
2. Coal Mines (undgd.)			
3. Coal (strip & auger)			
4. Petroleum & natural gas			
5. All other mining			
6. Building contractors			
7. Non-building contractors	.000403	.000403	0
8. Special contractors	.001058	.001058	0
9. Food products (n.e.c.)	.008346	.008343	-.000003
10. Dairies	.014273	.014267	-.000006
11. Bakeries			
12. Beverages			
13. Apparel & accessories			
14. Logging & sawmills			
15. Furn. & wood fabrication			
16. Printing & publishing	.001331	.013300	-.000001
17. Chemicals	0	.000030	.000030
18. Petroleum products			
19. Glass			
20. Stone & clay products			
21. Primary metals			
22. Fabricated metals			
23. Machinery (except electric)	.000403	.000403	0
24. Electric machinery			
25. Transportation equipment			
26. Instruments & products			
27. All other manufacturing	.001724	.001723	-.000001
28. Restaurants & bars	.001502	.001501	-.000001
29. Wholesale trade	.006673	.006670	-.000003
30. Retail food stores			
31. Auto service stations	.001512	.001511	-.000001
32. All other retail	.002147	.002146	-.000001
33. Banking	.000655	.000695	.000040
34. Other finance			
35. Insurance agents	.004163	.004161	-.000002
36. Real estate	.002429	.002428	-.000001
37. All other FIRE	.000857	.000856	-.000001
38. Hotels & lodgings			
39. Medical & legal services	.000534	.000534	0
40. Educational services			
41. All other services	.035884	.035870	-.000014
42. Railroads			
43. Trucking & warehousing	.004657	.004655	-.000002
44. All other transportation	.000060	.000060	0
45. Communications	.002036	.002035	-.000001
46. Electrical systems	.004506	.004514	.000008
47. Gas systems	.001018	.001018	0
48. Water & sanitary services	.001371	.001370	-.000001
49. Households	.183370	.183370	0
50. Out-of-state (imports)	.374830	.374933	.000103
51. All other final payments	.049440	.049452	.000012
52. Total	1.000000	1.000050	.000050

Table 3A-3 (cont.)

	Bakeries (11)		
	Original a_{ij}	New a_{ij}	Difference
1. Agriculture	.002338	.002337	-.000001
2. Coal Mines (undgd.)			
3. Coal (strip & auger)			
4. Petroleum & natural gas			
5. All other mining			
6. Building contractors	.000097	.000097	0
7. Non-building contractors			
8. Special contractors			
9. Food products (n.e.c.)	.001032	.001031	-.000001
10. Dairies	.007400	.007397	-.000003
11. Bakeries	.001048	.001048	0
12. Beverages			
13. Apparel & accessories			
14. Logging & sawmills			
15. Furn. & wood fabrication			
16. Printing & publishing	.001193	.001193	0
17. Chemicals		.000048	.000048
18. Petroleum products			
19. Glass			
20. Stone & clay products			
21. Primary metals			
22. Fabricated metals			
23. Machinery (except electric)			
24. Electric machinery			
25. Transportation equipment			
26. Instruments & products			
27. All other manufacturing	.005691	.005689	-.000002
28. Restaurants & bars	.000032	.000032	
29. Wholesale trade	.022749	.022739	-.000010
30. Retail food stores			
31. Auto service stations	.005707	.005705	-.000002
32. All other retail	.011254	.011249	-.000005
33. Banking	.006546	.006543	-.000003
34. Other finance	.003982	.003981	-.000001
35. Insurance agents	.021766	.021756	-.000010
36. Real estate			
37. All other FIRE	.000193	.000193	0
38. Hotels & lodgings	.000016	.000016	0
39. Medical & legal services	.000355	.000355	0
40. Educational services			
41. All other services	.045353	.045333	-.000020
42. Railroads			
43. Trucking & warehousing	.001870	.001869	-.000001
44. All other transportation	.000032	.000032	0
45. Communications	.005321	.005318	-.000003
46. Electrical systems	.052222	.052214	-.000008
47. Gas systems	.029231	.029217	-.000014
48. Water & sanitary services	.001870	.001869	-.000001
49. Households	.296230	.296235	.000005
50. Out-of-state (imports)	.430870	.431025	.000155
51. All other final payments	.045550	.045526	-.000024
52. Total	1.000000	1.000096	.000096

Table 3A-3 (cont.)

	Beverages (12)		
	Original a_{ij}	New a_{ij}	Difference
1. Agriculture			
2. Coal Mines (undgd.)	.000970	.000970	0
3. Coal (strip & auger)			
4. Petroleum & natural gas			
5. All other mining			
6. Building contractors			
7. Non-building contractors			
8. Special contractors			
9. Food products (n.e.c.)			
10. Dairies			
11. Bakeries			
12. Beverages	.007355	.007348	-.000007
13. Apparel & accessories	.002247	.002245	-.000002
14. Logging & sawmills			
15. Furn. & wood fabrication			
16. Printing & publishing	.014634	.014619	-.000015
17. Chemicals	.005006	.005077	-.000071
18. Petroleum products			
19. Glass	.013561	.013548	-.000013
20. Stone & clay products			
21. Primary metals	.008504	.008496	-.000008
22. Fabricated metals			
23. Machinery (except electric)			
24. Electric machinery			
25. Transportation equipment			
26. Instruments & products			
27. All other manufacturing	.010982	.010971	-.000011
28. Restaurants	.003014	.003011	-.000003
29. Wholesale trade	.012923	.012910	-.000013
30. Retail food stores			
31. Auto service stations	.009679	.009670	-.000009
32. All other retail	.001532	.001531	-.000001
33. Banking	.000409	.000510	.000101
34. Other finance			
35. Insurance agents	.012029	.012017	-.000012
36. Real estate			
37. All other FIRE			
38. Hotels & lodgings	.003014	.003011	-.000003
39. Medical & legal services	.002784	.002781	-.000003
40. Educational services			
41. All other services	.010982	.010971	-.000011
42. Railroads			
43. Trucking & warehousing	.017315	.017298	-.000017
44. All other transportation			
45. Communications	.012003	.011991	-.000012
46. Electrical systems	.005184	.005205	.000021
47. Gas systems	.001711	.001709	-.000002
48. Water & sanitary services	.005031	.005026	-.000005
49. Households	.257750	.257660	-.000090
50. Out-of-state (imports)	.326810	.327057	.000247
51. All other final payments	.254610	.254445	-.000165
52. Total	1.000000	1.000127	.000127

Table 3A-3 (cont.)

	Chemicals (17)		
	Original a_{ij}	New a_{ij}	Difference
1. Agriculture	.000010	.000010	0
2. Coal Mines (undgd.)	.010392	.011766	.001374
3. Coal (strip & auger)			
4. Petroleum & natural gas	.007533	.007526	-.000007
5. All other mining	.004321	.004317	-.000004
6. Building contractors	.000846	.000871	.000025
7. Non-building contractors	.001245	.001244	-.000001
8. Special contractors	.005750	.000594	-.005156
9. Food products (n.e.c.)	.000402	.000402	0
10. Dairies			
11. Bakeries			
12. Beverages			
13. Apparel & accessories	.000073	.000073	0
14. Logging & sawmills			
15. Furn. & wood fabrication	.000108	.000108	0
16. Printing & publishing	.001200	.001199	-.000001
17. Chemicals	.146151	.146561	.000410
18. Petroleum products	.001043	.001042	-.000001
19. Glass	.000622	.000621	-.000001
20. Stone & clay products	.001132	.001131	-.000001
21. Primary metals	.008294	.008318	.000024
22. Fabricated metals	.000677	.000677	0
23. Machinery (except electric)			
24. Electric machinery	.000945	.000944	-.000001
25. Transportation equipment			
26. Instruments & products	.001706	.001704	-.000002
27. All other manufacturing	.007466	.007486	.000020
28. Restaurants & bars	.000327	.000327	0
29. Wholesale trade	.013666	.013706	.000040
30. Retail food stores	.000104	.000104	0
31. Auto service stations	.000038	.000038	0
32. All other retail	.001527	.001526	-.000001
33. Banking	.000402	.000792	.000390
34. Other finance			
35. Insurance agents	.003106	.003103	-.000003
36. Real estate			
37. All other FIRE	.001161	.001160	-.000001
38. Hotels & lodgings	.000731	.000730	-.000001
39. Medical & legal services	.001407	.001406	-.000001
40. Educational services	.000156	.000156	0
41. All other services	.002722	.002719	-.000003
42. Railroads	.005646	.005661	.000015
43. Trucking & warehousing	.013383	.013423	.000040
44. All other transportation	.003504	.003501	-.000003
45. Communications	.003701	.003698	-.000003
46. Electrical systems	.009747	.010078	.000331
47. Gas systems	.015763	.015749	-.000014
48. Water & sanitary services	.001116	.001115	-.000001
49. Households	.156220	.156080	-.000140
50. Out-of-state (imports)	.372450	.374719	.002269
51. All other final payments	.193200	.193326	.000126
52. Total	1.000000	1.005365	.005365

Table 3A-3 (cont.)

	Petroleum products (18)		
	Original a_{ij}	New a_{ij}	Difference
1. Agriculture			
2. Coal Mines (undgd.)	.023277	.023355	.000078
3. Coal (strip & auger)			
4. Petroleum & natural gas	.006461	.066635	.000174
5. All other mining	.073867	.074061	.000194
6. Building contractors		.000009	.000009
7. Non-building contractors			
8. Special contractors	0	.000009	.000009
9. Food products (n.e.c.)			
10. Dairies			
11. Bakeries			
12. Beverages			
13. Apparel & accessories			
14. Logging & sawmills	.000785	.000787	.000002
15. Furn. & wood fabrication			
16. Printing & publishing	.001246	.001249	.000003
17. Chemicals	.045462	.045718	.000256
18. Petroleum products	.027646	.027718	.000072
19. Glass			
20. Stone & clay products	.006135	.006151	.000016
21. Primary metals		.000009	.000009
22. Fabricated metals	.006954	.006972	.000018
23. Machinery (except electric)			
24. Electric machinery			
25. Transportation equipment			
26. Instruments & products			
27. All other manufacturing	0	.000009	.000009
28. Restaurants & bars	.000299	.000299	0
29. Wholesale trade	.004642	.004671	.000029
30. Retail food stores			
31. Auto service stations			
32. All other retail	.000427	.000428	.000001
33. Banking	.006289	.006459	.000170
34. Other finance	.001203	.001206	.000003
35. Insurance agents	.000930	.000933	.000003
36. Real estate			
37. All other FIRE			
38. Hotels & lodgings			
39. Medical & legal services	.003891	.003901	.000010
40. Educational services			
41. All other services	.002466	.002472	.000006
42. Railroads	0	.000009	.000009
43. Trucking & warehousing	.015743	.015801	.000058
44. All other transportation	.015700	.015741	.000041
45. Communications	.020547	.020610	.000054
46. Electrical systems	.007475	.007520	.000045
47. Gas systems	.016707	.016751	.000044
48. Water & sanitary services	.000452	.000453	.000001
49. Households	.110450	.110002	-.000448
50. Out-of-state (imports)	.376310	.378270	.001960
51. All other final payments	.164650	.165215	.000565
52. Total	1.000000	1.004432	.004432

Table 3A-3 (cont.)

	Glass (19)		
	Original a_{ij}	New a_{ij}	Difference
1. Agriculture			
2. Coal Mines (undgd.)	.000239	.000259	.000020
3. Coal (strip & auger)	.000064	.000064	0
4. Petroleum & natural gas			
5. All other mining	.018071	.018022	-.000049
6. Building contractors	0	.000015	.000015
7. Non-building contractors			
8. Special contractors	.000108	.000123	.000015
9. Food products (n.e.c.)			
10. Dairies			
11. Bakeries			
12. Beverages			
13. Apparel & accessories			
14. Logging & sawmills			
15. Furn. & wood fabrication	.006319	.006302	-.000017
16. Printing & publishing	.003485	.003476	-.000009
17. Chemicals	.004378	.004576	.000198
18. Petroleum products			
19. Glass	.035597	.035501	-.000096
20. Stone & clay products	.003059	.003051	-.000008
21. Primary metals	.002700	.002713	-.000013
22. Fabricated metals	.000095	.000095	0
23. Machinery (except electric)			
24. Electric machinery			
25. Transportation equipment			
26. Instruments & products			
27. All other manufacturing	.049471	.049353	-.000118
28. Restaurants & bars			
29. Wholesale trade	.006480	.006474	-.000006
30. Retail food stores			
31. Auto service stations	.000092	.000092	0
32. All other retail	.002754	.002746	-.000008
33. Banking	.004409	.004628	.000219
34. Other finance			
35. Insurance agents	.019282	.019230	-.000052
36. Real estate	.000246	.000246	0
37. All other FIRE	.000423	.000422	-.000001
38. Hotels & lodgings	.000121	.000120	-.000001
39. Medical & legal services	.000585	.000584	-.000001
40. Educational services	.000064	.000064	0
41. All other services	.029299	.029220	-.000079
42. Railroads	0	.000015	.000015
43. Trucking & warehousing	.003762	.003783	.000021
44. All other transportation			
45. Communications	.003390	.003381	-.000009
46. Electrical systems	.038839	.038774	-.000065
47. Gas systems	.024595	.024528	-.000067
48. Water & sanitary services	.000295	.000294	-.000001
49. Households	.303420	.303000	-.000420
50. Out-of-state (imports)	.179980	.180954	.000974
51. All other final payments	.258370	.257876	-.000494
52. Total	.999992	.999992	0

Table 3A-3 (cont.)

	Stone & clay products (20)		
	Original a_{ij}	New a_{ij}	Difference
1. Agriculture	.002487	.002480	-.000007
2. Coal Mines (undgd.)	.008909	.008907	-.000002
3. Coal (strip auger)			
4. Petroleum & natural gas			
5. All other mining	.041557	.041440	-.000117
6. Building contractors		.000078	.000018
7. Non-building contractors			
8. Special contractors	.000533	.000550	.000017
9. Food products (n.e.c.)			
10. Dairies			
11. Bakeries			
12. Beverages			
13. Apparel & accessories	.000480	.000478	-.000002
14. Logging & sawmills	.000058	.000058	0
15. Furn. & wood fabrication			
16. Printing & publishing	.006341	.006324	-.000017
17. Chemicals	.004845	.005068	.000223
18. Petroleum products	.037654	.037548	-.000106
19. Glass			
20. Stone & clay products	.021171	.021111	-.000060
21. Primary metals	0	.000022	.000022
22. Fabricated metals	.036426	.036323	-.000103
23. Machinery (except electric)	.000717	.000715	-.000002
24. Electric machinery	.000112	.000112	0
25. Transportation equipment	.001120	.001117	-.000003
26. Instruments & products			
27. All other manufacturing	.004499	.004505	.000006
28. Restaurants & bars			
29. Wholesale trade	.012862	.012862	0
30. Retail food stores			
31. Auto service stations	.003684	.003673	-.000011
32. All other retail	.018253	.018202	-.000051
33. Banking	.000609	.000867	.000258
34. Other finance			
35. Insurance agents	.008233	.008209	-.000024
36. Real estate			
37. All other FIRE			
38. Hotels & lodgings	.000018	.000018	0
39. Medical & legal services	.006821	.006802	-.000019
40. Educational services			
41. All other services	.015560	.015516	-.000044
42. Railroads	.002398	.002409	.000011
43. Trucking & warehousing	.034561	.034500	-.000061
44. All other transportation	.000323	.000322	-.000001
45. Communications	.006902	.006882	-.000020
46. Electrical systems	.031245	.031202	-.000043
47. Gas systems	.034063	.034965	-.000098
48. Water & sanitary services	.000609	.000608	-.000001
49. Households	.294200	.293822	-.000378
50. Out-of-state (imports)	.175330	.176482	.001152
51. All other final payments	.186440	.186112	-.000328
52. Total	1.000000	1.000215	.000215

Table 3A-3 (cont.)

	Primary metal products (21)		
	Original a_{ij}	New a_{ij}	Difference
1. Agriculture	.001781	.001760	-.000021
2. Coal Mines (undgd.)	.019301	.019179	-.000122
3. Coal (strip & auger)			
4. Petroleum & natural gas	.000197	.000194	-.000003
5. All other mining	.004606	.004550	-.000056
6. Building contractors	.000022	.000040	.000018
7. Non-building contractors	.000044	.000043	-.000001
8. Special contractors	.001883	.001945	.000062
9. Food products (n.e.c.)			
10. Dairies			
11. Bakeries			
12. Beverages			
13. Apparel & accessories	.000007	.000007	0
14. Logging & sawmills	.000007	.000007	0
15. Furn. & wood fabrication			
16. Printing & publishing	.000128	.000127	-.000001
17. Chemicals	.003803	.003784	-.000019
18. Petroleum products	.000102	.000101	-.000001
19. Glass	.000279	.000276	-.000003
20. Stone & clay products			
21. Primary metals	.003976	.003956	-.000020
22. Fabricated metals	.001527	.001508	-.000019
23. Machinery (except electric)	.000457	.000451	-.000006
24. Electric machinery	.000380	.000397	.000017
25. Transportation equipment	.000068	.000067	-.000001
26. Instruments & products			
27. All other manufacturing	.001022	.001028	.000006
28. Restaurants & bars	.000043	.000042	-.000001
29. Wholesale trade	.004786	.004832	.000046
30. Retail food stores			
31. Auto service stations	.000003	.000003	0
32. All other retail	.002533	.002502	-.000031
33. Banking	.000980	.001988	.001008
34. Other finance	.000165	.000163	-.000002
35. Insurance agents	.000760	.000751	-.000009
36. Real estate			
37. All other FIRE	.000633	.000625	-.000008
38. Hotels & lodgings	.000189	.000186	-.000003
39. Medical & legal services	.001164	.001150	-.000014
40. Educational services	.000398	.000393	-.000005
41. All other services	.007939	.007843	-.000096
42. Railroads	.008492	.008436	-.000056
43. Trucking & warehousing	.008236	.008239	.000003
44. All other transportation	.009152	.009041	-.000111
45. Communications	.002563	.002532	-.000031
46. Electrical systems	.007231	.007284	.000053
47. Gas systems	.002317	.003513	.001196
48. Water & sanitary services	.000565	.000558	-.000007
49. Households	.234490	.233182	-.001308
50. Out-of-state (imports)	.567890	.568077	.000187
51. All other final payments	.099880	.099239	-.000064
52. Total	.999999	.999279	-.000720

Table 3A-3 (cont.)

	Fabricated metals (22)		
	Original a_{ij}	New a_{ij}	Difference
1. Agriculture	.000010	.000010	0
2. Coal Mines (undgd.)	.000246	.000275	.000029
3. Coal (strip & auger)			
4. Petroleum & natural gas			
5. All other mining	.000340	.000339	-.000001
6. Building contractors	0	.000020	.000020
7. Non-building contractors			
8. Special contractors	.001513	.001527	.000014
9. Food products (n.e.c.)			
10. Dairies			
11. Bakeries			
12. Beverages			
13. Apparel & accessories			
14. Logging & sawmills			
15. Furn. & wood fabrication			
16. Printing & publishing	.001996	.001989	-.000007
17. Chemicals	.001794	.002073	.000279
18. Petroleum products			
19. Glass	.000015	.000015	0
20. Stone & clay products			
21. Primary metals	.117139	.116771	-.000368
22. Fabricated metals	.000867	.000864	-.000003
23. Machinery (except electric)	.000237	.000236	-.000001
24. Electric machinery	.000030	.000029	-.000001
25. Transportation equipment			
26. Instruments & products	.000237	.000236	-.000001
27. All other manufacturing	.005874	.005874	0
28. Restaurants & bars	.000384	.000383	-.000001
29. Wholesale trade	.011235	.011237	.000002
30. Retail food stores	.000010	.000010	0
31. Auto service stations	.000128	.000128	0
32. All other retail	.003405	.003394	-.000011
33. Banking	.000172	.000481	.000309
34. Other finance	.000301	.000300	-.000001
35. Insurance agents	.002878	.002868	-.000010
36. Real estate	.005071	.005054	-.000017
37. All other FIRE	.001370	.001365	-.000005
38. Hotels & lodgings	.000123	.000123	0
39. Medical & legal services	.000237	.000236	-.000001
40. Educational services	.000030	.000029	-.000001
41. All other services	.003292	.003281	-.000011
42. Railroads	0	.000020	.000020
43. Trucking & warehousing	.002124	.002156	.000032
44. All other transportation	.000064	.000064	0
45. Communications	.001754	.001748	-.000006
46. Electrical systems	.009624	.009646	.000022
47. Gas systems	.006943	.006920	-.000023
48. Water & sanitary services	.002454	.002446	-.000008
49. Households	.289600	.289171	-.000429
50. Out-of-state (imports)	.433560	.434064	.000504
51. All other final payments	.094940	.094891	-.000049
52. Total	1.000000	1.000275	.000275

Table 3A-3 (cont.)

	Machinery (except electric) (23)		
	Original a_{ij}	New a_{ij}	Difference
1. Agriculture			
2. Coal Mines (undgd.)	.000089	.000100	.000011
3. Coal (strip & auger)			
4. Petroleum & natural gas			
5. All other mining			
6. Building contractors		.000012	.000012
7. Non-building contractors			
8. Special contractors	.000219	.000230	.000011
9. Food products (n.e.c.)			
10. Dairies			
11. Bakeries			
12. Beverages			
13. Apparel & accessories			
14. Logging & sawmills	.000095	.000094	-.000001
15. Furn. & wood fabrication			
16. Printing & publishing	.006873	.006860	-.000013
17. Chemicals		.000148	.000148
18. Petroleum products			
19. Glass			
20. Stone & clay products			
21. Primary metals	.038297	.038240	-.000057
22. Fabricated metals	.024682	.024637	-.000045
23. Machinery (except electric)	.000018	.000018	0
24. Electric machinery	.013320	.013296	-.000024
25. Transportation equipment			
26. Instruments & products			
27. All other manufacturing	.003750	.003755	.000005
28. Restaurants & bars	.000243	.000242	-.000001
29. Wholesale trade	.009310	.009316	.000006
30. Retail food stores			
31. Auto service stations	.000313	.000303	0
32. All other retail	.000597	.000596	-.000001
33. Banking	.002460	.002657	.000197
34. Other finance			
35. Insurance agents	.009534	.009517	-.000017
36. Real estate	.000130	.000130	0
37. All other FIRE			
38. Hotels & lodgings	.000367	.000366	-.000001
39. Medical & legal services	.000272	.000272	0
40. Educational services			
41. All other services	.002259	.002255	-.000004
42. Railroads	.002277	.002273	-.000004
43. Trucking & warehousing	.015047	.015020	-.000027
44. All other transportation	.007458	.007445	-.000013
45. Communications	.001076	.001075	-.000001
46. Electrical systems	.006613	.006601	-.000012
47. Gas systems	.002307	.002303	-.000004
48. Water & sanitary services	.000172	.000171	-.000001
49. Households	.274430	.274213	.000217
50. Out-of-state (imports)	.441750	.442056	.000306
51. All other final payments	.136020	.135932	-.000088
52. Total	1.000000	1.000201	.000201

Table 3A-3 (cont.)

	Electrical machinery (24)		
	Original a_{ij}	New a_{ij}	Difference
1. Agriculture			
2. Coal Mines (undgd.)	0	.000004	.000004
3. Coal (strip & auger)			
4. Petroleum & natural gas			
5. All other mining			
6. Building contractors	0	.000004	.000004
7. Non-building contractors	.000947	.000743	-.000204
8. Special contractors	.000033	.000030	-.000003
9. Food products (n.e.c.)			
10. Dairies			
11. Bakeries			
12. Beverages			
13. Apparel & accessories			
14. Logging & sawmills			
15. Furn. & wood fabrication	.000230	.000180	-.000050
16. Printing & publishing	.001122	.000880	-.000242
17. Chemicals	.024236	.019068	-.005168
18. Petroleum products			
19. Glass			
20. Stone & clay products			
21. Primary metals	.157560	.123632	-.033928
22. Fabricated metals			
23. Machinery (except electric)			
24. Electric machinery	.001023	.0C0803	-.000220
25. Transportation equipment			
26. Instruments & products	.002873	.002254	-.000619
27. All other manufacturing	.000542	.000409	-.000113
28. Restaurants & bars	.000257	.000202	-.000055
29. Wholesale trade	.003590	.002825	-.000765
30. Retail food stores			
31. Auto service stations	.000115	.000090	-.000025
32. All other retail	.003869	.003036	-.000833
33. Banking	.000985	.000846	-.000139
34. Other finance			
35. Insurance agents	007223	.005668	-.001555
36. Real esate	.001297	.001018	-.000279
37. All other FIRE			
38. Hotels & lodgings	.000930	.000730	-.000200
39. Medical & legal services	.001664	.001305	-.000359
40. Educational services	.004339	.003405	-.000934
41. All other services	.011070	.008686	-.002384
42. Railroads	.003338	.002623	-.000715
43. Trucking & warehousing	.007677	.006038	-.001644
44. All other transportation	.027996	.021967	-.006029
45. Communications	.008487	.006660	-.001827
46. Electrical systems	.032401	.240116	.207715
47. Gas systems	.009899	.007767	-.002132
48. Water & sanitary services	.004558	.003577	-.000981
49. Households	.271640	.271536	-.000104
50. Out-of-state (imports)	.267420	.267693	.000273
51. All other final payments	.142670	.142624	-.000046
52. Total	1.000000	1.000066	.000066

Table 3A-3 (cont.)

	Transportation Equipment (25)		
	Original a_{ij}	New a_{ij}	Difference
1. Agriculture			
2. Coal Mines (undgd.)	0	.000013	.000013
3. Coal (strip & auger)			
4. Petroleum & natural gas			
5. All other mining	.002656	.002640	-.000016
6. Building contractors	0	.000008	.000008
7. Non-building contractors			
8. Special contractors	.000021	.000030	.000009
9. Food products (n.e.c.)			
10. Dairies			
11. Bekeries			
12. Beverages			
13. Apparel & accessories			
14. Logging & sawmills			
15. Furn. & wood fabrication			
16. Printing & publishing	.000183	.000182	-.000001
17. Chemicals	0	.000123	.000123
18. Petroleum products			
19. Glass			
20. Stone & clay products			
21. Primary metals	.072342	.071938	-.000404
22. Fabricated metals			
23. Machinery (except electric)			
24. Electric machinery			
25. Transportation equipment			
26. Instruments & products			
27. All other manufacturing	.000136	.000144	.000008
28. Restaurants & bars	.000089	.000089	0
29. Wholesale trade	.002307	.002310	.000003
30. Retail food stores			
31. Auto service stations	.000009	.000008	-.000001
32. All other retail	.000749	.000745	-.000004
33. Banking	.001017	.001176	.000159
34. Other finance			
35. Insurance agents	.001732	.001722	-.000010
36. Real estate			
37. All other FIRE			
38. Hotels & lodgings	.000034	.000034	0
39. Medical & legal services	.001060	.001054	-.000006
40. Educational services			
41. All other services	.005137	.005107	-.000030
42. Railroads	.019560	.019456	-.000104
43. Trucking & warehousing	.000651	.000664	.000013
44. All other transportation			
45. Communications	.007907	.007862	-.000045
46. Electrical systems	.008597	.012804	.004207
47. Gas systems	.005009	.004980	-.000029
48. Water & sanitary services	.002047	.002035	-.000012
49. Households	.255040	.253800	-.001240
50. Out-of-state (imports)	.489880	.487953	-.001927
51. All other final payments	.123830	.123240	-.000590
52. Total	1.000000	1.000110	.000110

Table 3A-3 (cont.)

	Instruments & products (26)		
	Original a_{ij}	New a_{ij}	Difference
1. Agriculture			
2. Coal Mines (undgd.)	0	.000145	.000145
3. Coal (strip & auger)			
4. Petroleum & natural gas			
5. All other mining			
6. Building contractors	0	.000145	.000145
7. Non-building contractors			
8. Special contractors	0	.000145	.000145
9. Food products (n.e.c.)			
10. Dairies			
11. Bakeries			
12. Beverages			
13. Apparel & accessories			
14. Logging & sawmills			
15. Furn. & wood fabrication			
16. Printing & publishing	.000443	.000434	-.000009
17. Chemicals	.059960	.060494	.000534
18. Petroleum products			
19. Glass	.003688	.003614	-.000074
20. Stone & clay products			
21. Primary metals	0	.000145	.000145
22. Fabricated metals			
23. Machinery (except electric)			
24. Electric machinery			
25. Transportation equipment			
26. Instruments & products			
27. All other manufacturing	.008186	.008167	-.000019
28. Restaurants & bars			
29. Wholesale trade	.011653	.001636	-.000017
30. Retail food stores			
31. Auto service stations			
32. All other retail			
33. Banking	.001475	.001662	.000187
34. Other finance			
35. Insurance agents	.001696	.001662	-.000034
36. Real estate			
37. All other FIRE			
38. Hotels & lodgings			
39. Medical & legal services	.001918	.001879	-.000039
40. Educational services			
41. All other services	.003171	.003138	.000038
42. Railroads	0	.000145	.000145
43. Trucking & warehousing	.005310	.005421	.000102
44. All other transportation			
45. Communications	.001549	.001518	-.000031
46. Electrical systems	.001623	.001951	.000328
47. Gas systems	.001033	.001012	-.000021
48. Water & sanitary services	.000221	.000217	-.000004
49. Households	.289590	.287005	-.002585
50. Out-of-state (imports)	.522310	.520382	-.001928
51. All other final payments	.086270	.086224	-.000046
52. Total	1.000096	.999141	-.000955

Table 3A-3 (cont.)

	Electrical systems (46)		
	Original a_{ij}	New a_{ij}	Difference
1. Agriculture			
2. Coal Mines (undgd.)	.089369	.087868	-.001501
3. Coal (strip & auger)	.033872	.033303	-.000569
4. Petroleum & natural gas			
5. All other mining			
6. Building contractors			
7. Non-building contractors			
8. Special contractors	.000011	.000011	0
9. Food products (n.e.c.)			
10. Dairies			
11. Bakeries			
12. Beverages			
13. Apparel & accessories			
14. Logging & sawmills			
15. Furn. & wood fabrication	.000199	.000195	-.000004
16. Printing & publishing	.003954	.003888	-.000066
17. Chemicals			
18. Petroleum products	.005274	.005185	-.000089
19. Glass			
20. Stone & clay products	.002595	.002552	-.000043
21. Primary metals	.000464	.000457	-.000007
22. Fabricated metals	.006239	.006201	-.000038
23. Machinery (except electric)			
24. Electric machinery	.000199	.000195	-.000004
25. Transportation equipment			
26. Instruments & products			
27. All other manufacturing	.000007	.000007	0
28. Restaurants & bars	.000620	.000609	-.000011
29. Wholesale trade	.003289	.003234	-.000055
30. Retail food stores			
31. Auto service stations	.000083	.000082	-.000001
32. All other retail	.000578	.000569	-.000009
33. Banking	.000051	.004657	.004606
34. Other finance			
35. Insurance agents	.000902	.000887	-.000015
36. Real estate	.001153	.001134	-.000019
37. All other FIRE	.001905	.001873	-.000032
38. Hotels & lodgings	.001153	.001134	-.000019
39. Medical & legal services	.001272	.001251	-.000021
40. Educational services	.000007	.000007	0
41. All other services	.003383	.003326	-.000057
42. Railroads	.013507	.013281	-.000226
43. Trucking & warehousing	.020820	.020806	-.000014
44. All other transportation	.003813	.003749	-.000064
45. Communications	.004869	.004787	-.000082
46. Electrical systems	.055628	.055869	.000241
47. Gas systems	.001101	.001082	-.000019
48. Water & sanitary services	.000070	.000069	-.000001
49. Households	.206190	.203567	-.002623
50. Out-of-state (imports)	.152810	.160010	.007200
51. All other final payments	.384620	.378156	-.006464
52. Total	1.000000	.999998	-.000002

4 The Control of Sulfur Dioxide Emissions

Industry Sulfur Emissions Standards

National primary air quality standards were set at levels designed to protect public health with allowances for reasonable margins of safety. The objective of secondary standards was to protect the public from further adverse effects, including such factors as soiling, corrosion, and vegetation damage. To meet these standards, sulfur oxide emissions in West Virginia are to be limited according to WVAPCC Regulation X. Each air pollution region in West Virginia is classified into one of the three priority categories for sulfur dioxide given in Table 4-1. All of the areas classified as Priority I and II regions are in northern West Virginia.

Regulation X limits emissions from the combustion of fossil fuels, from sulfuric acid plants, and from other sources. The major causes of SO_2 emissions in West Virginia are coal combustion or off-gases generated by process operations using coal. These are the only sources of SO_2 emissions considered in the present study. The allowable emissions from fuel combustion vary from region to region within the state (see Figure 1-1). By July 1, 1975, the regulations call for emissions from electric utilities consistent with coal sulfur content of 1.6 percent in Priority Regions I and II, and 2.0 percent in Priority Region III. In the case of industrial plants, a sulfur content of approximately 2.0 percent will be acceptable in all three priority regions. Regulations already in force call for 1 percent sulfur levels in the Kanawha Valley region. By 1978, the corresponding levels of sulfur content in coal in all priority regions, except the Kanawha Valley region, are expected to be reduced to approximately 1.25 percent for electric utilities and 1.4 percent for industrial plants.[1]

Affected Industries

The major West Virginia industrial sectors with SO_2 emission problems are electric utilities, chemicals, and primary metals. Other industries do not use large quantities of high-sulfur fuel, and the direct costs of sulfur abatement in sectors other than those mentioned above will not be considered here.

Table 4-1
Ambient Sulfur Dioxide Concentrations (Micrograms per Cubic Meter)

	Priority I	Priority II	Priority III	Primary Standards	Secondary Standards
Annual Arithmetic Mean	100	60-100	60	80	60
24-hr. Maximum	455	260-455	260	365	260
3-hr. Maximum		1300	1300		1300

Electric Utilities and Systems

About 99 percent of the electric power generated in West Virginia is obtained by the combustion of coal, so the SO_2 regulations are vital to this sector. In the southern part of the state, most power plants have adequate low-sulfur coal supplies. With one exception, the primary regulation does not directly affect plants located there.[2] The Cabin Creek plant, located near Charleston, is expected to close in 1975 because the investment required to retrofit for particulate abatement does not seem to be justified at present. Nine northern West Virginia plants are affected by the regulations, and the Windsor plant located between Wheeling and Weirton has been closed because of both sulfur dioxide and particulate regulations.

Chemicals

Most of the emissions in this sector come from coal combustion in steam-generating boilers. The chemical plants in the northern counties will bear the brunt of meeting Regulation X. There are a few sulfuric acid plants in West Virginia, but their problem is primarily one of acid mist. This is covered by particulate regulations and was discussed in the last chapter. Small gas flares in some chemical plants also emit sulfur pollutants, but these are negligible and the Commission generally has allowed variances.

Primary Metal Products

Coal is the source of sulfur oxide pollutants that must be restricted. But the actual emissions are caused by the volatilization of coal to coke while producing coke-oven gas and by-product coke breeze. There are also some emissions from direct coal combustion for steam generation.

Compliance Methods to Limit Sulfur Oxide Emissions

A number of alternate strategies are available for the control of SO_2 emissions. The most common strategy to meet primary standards (1975 regulations) is some form of fuel-switching or fuel desulfurization. For large coal users, the usual strategy is to switch from a high-sulfur to a low-sulfur coal. Most electric utilities and large chemical plants plan to use this strategy to meet primary standards. Fuel-switching may require that some coal be imported from other states. Where this strategy has been followed, it generally has resulted in the blending of high-sulfur West Virginia coal and imported low-sulfur coal. Although this is the simplest strategy available, it usually requires investment in new facilities for blending. Also, fuel-switching cannot be considered as an acceptable *long-run* solution to the problem of sulfur emissions because of the relative scarcity of low-sulfur coal.

Some chemical and metallurgical concerns were switching from coal to oil at the time of this study. This type of switch requires modification of the boiler to allow liquid fuel injection ports. Storage and fuel-handling facilities for oil also must be installed. This is not likely to be a major method of compliance with sulfur regulations, however, in view of events following renewal of the Mideast war late in 1973. Another method of compliance is to add crushed lime-based materials to pulverized coal. This causes a chemical reaction that permits capture by slag of the SO_2 in the flue gas. This method will work only where a small decrease in SO_2 emissions is required to meet Regulation X.

The only completely reliable method for removing sulfur oxides in large-scale plants at present is that of fuel-switching. But since this is not an acceptable long-run solution to the problem of sulfur emissions, we have considered three alternative methods in this study.

The first method—*physical coal-cleaning*—is relatively inexpensive but also relatively ineffective. It is, however, technically feasible at the present time. Sulfur is found in coal in two principal forms, organic and pyritic. The organic sulfur is chemically bound to the coal matrix in a complex way and cannot be removed by physical cleaning. Pyritic sulfur is not bound to the coal matrix and can be removed by physical means. Physical or "deep" coal-cleaning can eliminate most of the pyritic sulfur embedded in coal from a typical Appalachian seam. Although a small amount of coal is lost in this process, it can reduce the sulfur content of the remaining coal up to 40 percent. A physical coal-cleaning process is described schematically in Figure 4-1.

A more effective approach to the reduction of the sulfur content of coal is *chemical cleaning*. Several options are available, but the one discussed here is called the *Myers-TRW* (Thompson-Ramo-Woolridge) method. This method uses

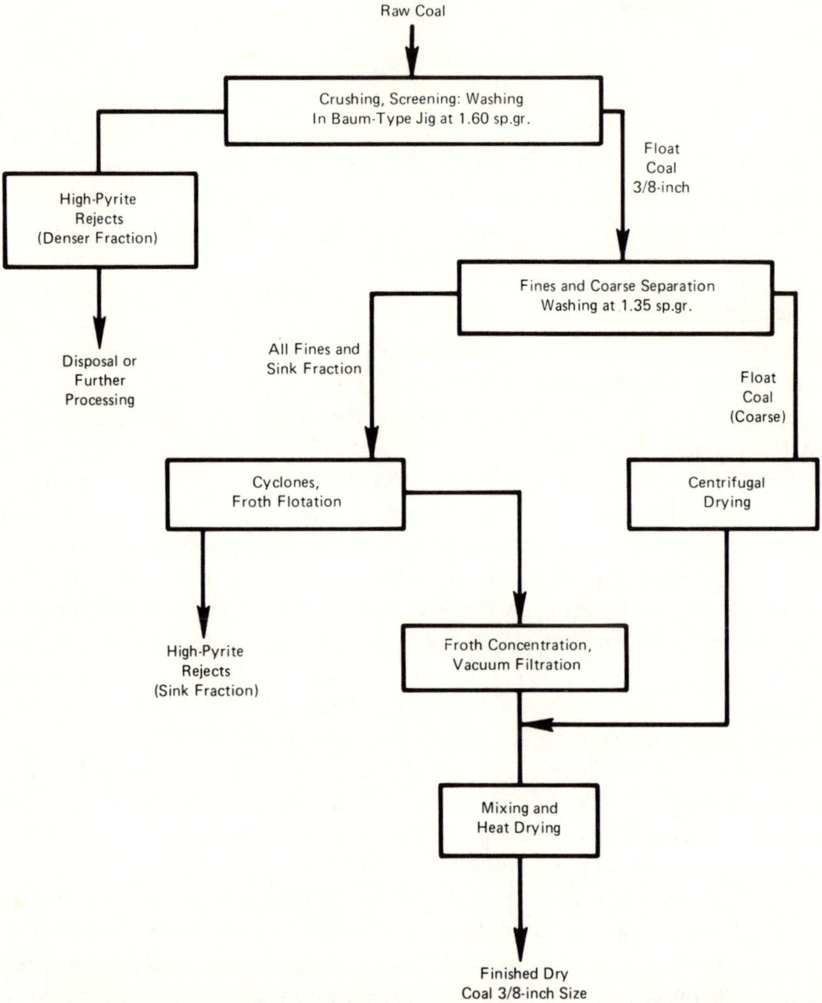

Figure 4-1. Schematic of Typical "Deep" Coal-Cleaning Operation

chemical leaching to remove pyritic sulfur from coal particles. The extracted pyrite is oxidized to obtain elemental sulfur. The free sulfur is then removed by solvent extraction, and the solvent (ferric sulfate solution) is recycled. A schematic of this process is given in Figure 4-2. Unfortunately, the process is not yet technically feasible for large-scale operations such as electric utilities. We have estimated the cost of adopting this method, however, since it might become technically feasible within a few years.

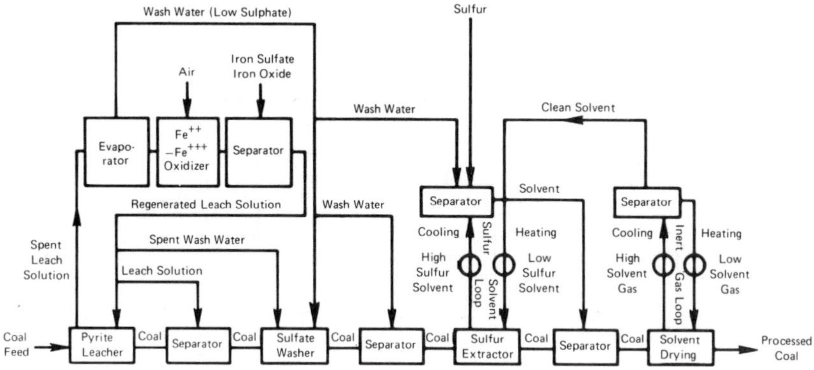

Figure 4-2. Schematic of Proposed Myers Process

The third method examined is SO_2 *flue-gas scrubbing.* Although we have estimated the cost of adopting this method, it also has not been completely accepted by the electric utilities as technically feasible. This is a complex operation from a chemical and engineering point of view, and no attempt will be made to describe it in detail here. Two variations of flue-gas scrubbing are presented schematically in Figure 4-3.

The Myers-TRW sulfur removal method, if perfected, could remove up to 95 percent of the pyritic sulfur in coal. This would be expected to reduce total sulfur dioxide emissions approximately 45 to 50 percent. The Myers-TRW method would be increasingly effective as a function of a declining organic/pyritic ratio.

Chemical cleaning of coal could be used in conjunction with flue-gas scrubbing, and this would be the most effective method for controlling sulfur dioxide emissions. It has been estimated that this combination would reduce emissions by 50 to 70 percent. If the design of future scrubbing systems can be improved, it is estimated that their efficiency will range between 80 to 90 percent. There are reasons to be optimistic about this possibility. Some pilot scrubbing operations have achieved 90 percent removal efficiency for over a year. In most instances, an 80 percent reduction in sulfur dioxide emissions would meet present secondary standards in West Virginia for Appalachian coal.

The three methods described above are not mutually exclusive alternatives at the sectoral level. Some establishments might be able to meet prescribed quality standards by adopting nothing more costly and effective than deep coal-cleaning; others might be forced to turn to one of the more expensive and more effective methods. To estimate costs, it was necessary to make rather precise assumptions, however. These are as follows:

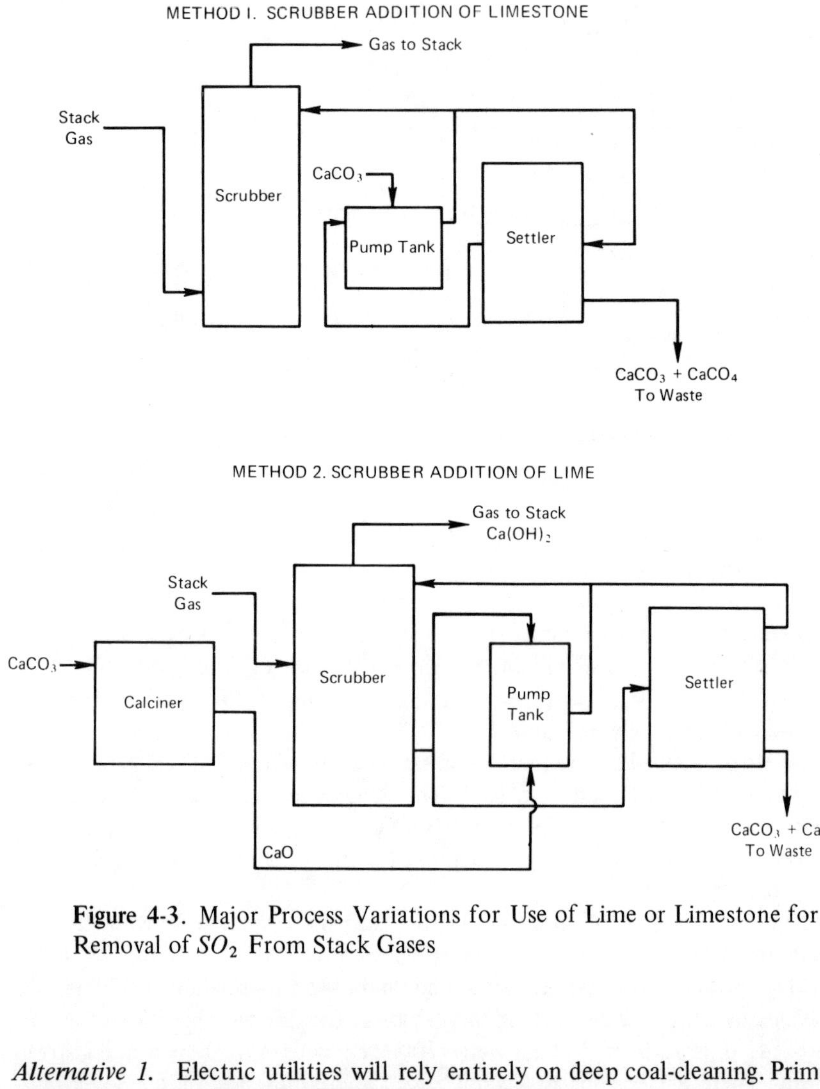

Figure 4-3. Major Process Variations for Use of Lime or Limestone for Removal of SO_2 From Stack Gases

Alternative 1. Electric utilities will rely entirely on deep coal-cleaning. Primary metals and chemicals will reduce sulfur emissions by a combination of fuel-switching and the installation of wet scrubbers.

Alternative 2. Utility plants representing 16 percent of the capacity of those plants now burning high-sulfur coal will use deep coal-cleaning; all remaining plants representing 84 percent of high-sulfur coal burning capacity will use the Myers–TRW method. The assumptions of Alternative 1 are unchanged for primary metals and chemicals.

Alternative 3. Utility plants representing 16 percent of capacity will use the Myers-TRW method; plants accounting for the remaining 84 percent of capacity will install flue-gas scrubbers. Capital spending will be at the rate of $96/kw of capacity.[3] The assumptions of Alternative 1 are unchanged for primary metals and chemicals.

Separate cost estimates have been made for each of the three alternatives. The relationship between cost and effectiveness of the alternatives is summarized in Figure 4-4. The horizontal axis of Figure 4-4 shows the percent of sulfur dioxide removal. The left-hand vertical axis shows the capital cost of the three alternatives expressed in dollars per kw of capacity. The righthand axis shows the cost of the three alternatives in terms of additional production required to produce a constant bill of goods for delivery to final demand, the measure of "real" cost used in this study. As Figure 4-4 shows, there is a wide range between the maximum and minimum estimates of the cost of flue-gas scrubbing processes for a given level of sulfur removal.

Estimated Engineering Costs of SO_2 Control

All major companies in West Virginia have certified compliance schedules, indicating the procedures by which they will meet the primary standards of WVAPCC Regulation X.[4] Using these schedules as a base, the cost of compliance was estimated for the chemical and primary metals sectors. Almost all of the strategies utilized will also allow compliance by these sectors with secondary standard regulations.

The electric utility strategies call for compliance with primary standards, but only two utility plants under the proposed strategies could meet secondary standard regulations. These strategies were formulated late in 1973. We estimated the costs of SO_2 reduction by chemical coal-cleaning and flue-gas desulfurization techniques to permit the comparison of costs for meeting secondary standard requirements.

For the chemical and metal industries, the cost of low-sulfur fuel was estimated at roughly twice the cost of high-sulfur coal on a comparable energy basis.[5] Boiler modification capital costs were estimated using standard costs to modify a unit designed for one type of fuel to accept new fuel and combustion techniques.[6] Cost estimates of using electricity for power instead of internal generation of power and steam by the combustion of fossil fuel were based on published bus-bar electricity and transformer costs. Scrubbing system costs were estimated primarily from EPA data.[7] The capital cost of a limestone scrubbing system was estimated at $96/kw of capacity.[8] The estimated capital cost of "deep" coal-cleaning is $8,000/ton coal-hour. A $6 million/650 thousand ton-year capacity was used as a base cost of the Myers (TRW) coal-cleaning process.[9] A 0.7 power factor was used for scale-up of this process, whereas a 1.0 power factor scale-up was used for deep coal-cleaning and scrubbing costs.

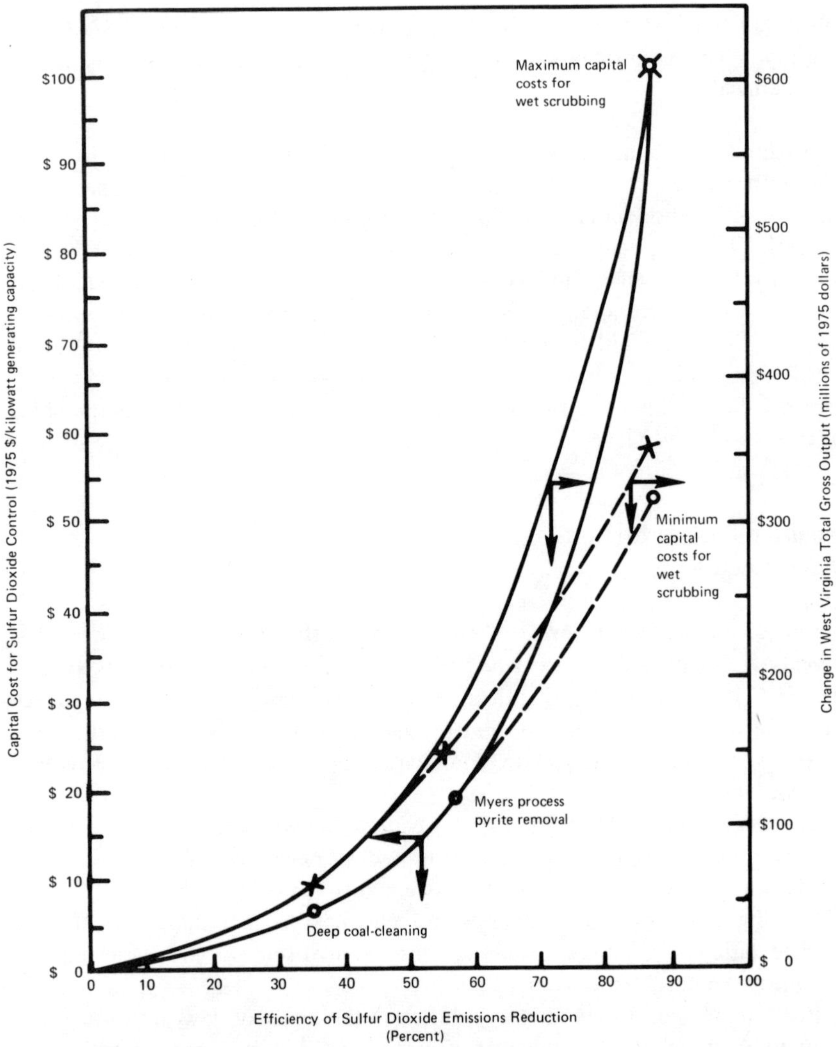

Figure 4-4. Cost of Sulfur Dioxide Control as a Function of Per Cent Reduction of Sulfur Dioxide Emissions at Given Electric Utility Sites.

For the electric utility sector, several strategies were examined. One strategy was to assume that deep coal-cleaning would be used by all plants, with aggregate capacity of 6,750 mw, to achieve a 30 percent reduction in SO_2 emissions. We estimated that the alternative Myers-TRW process would achieve 50 to 60 percent reductions. Finally, sulfur dioxide scrubbing in plants accounting for 5,600 mw of capacity, plus coal cleaning in plants accounting for 1,150 mw of capacity, could be used to reduce emissions by about 85 percent and thus meet secondary standards.

The Direct Cost of Sulfur Dioxide Abatement

The direct capital and operating cost of sulfur dioxide abatement under the three alternative systems discussed above are given in Table 4-2. This table shows graphically that under any of the alternatives most of the cost of sulfur dioxide abatement falls on the electric utilities. If we assume that utilities rely entirely on physical coal-cleaning, while primary metals and chemicals rely on fuel-switching and the use of wet scrubbers, we estimate the capital cost incurred by the three major polluters at \$42.6 million. Ninety-five percent of this would be charged to electric utilities. The direct annual operating costs of maintaining deep coal-cleaning and wet scrubbing facilities in these three sectors would amount to about \$17.5 million, and 75 percent of this total would be charged to electric utilities.

Under Alternative 2, the initial capital outlays of utilities would rise from \$40.5 million to \$127.4 million, an increase of 216 percent. Annual operating costs in electric utilties would also go uo from \$13 million to more than \$36 million, or 177 percent. Since the cost of abatement in chemicals and primary metals would remain unchanged, the share of total sulfur dioxide abatement cost charged to electric utilities would increase.

There would be a quantum jump if electric utilities adopted Alternative 3, the combination of Myers-TRW chemical coal-cleaning and SO_2 flue-gas scrubbing. In this event, the initial capital cost—at the rate of \$96/kw—would amount to \$664.3 million, or 422 percent more than Alternative 2. Annual operating costs would also go up to \$109 million, or an increase over Alternative 2 of slightly more than 200 percent. Since we have assumed the same abatement strategy in chemicals and primary metals under all three alternatives, while we have assumed that electric utilities would adopt increasingly costly methods of control, the share of the total sulfur dioxide abatement bill charged to electric utilities increases substantially as we move from Alternative 1 to Alternative 2, and thence to Alternative 3.

The changes in total capital and operating coefficients due to the installation and operation of sulfur dioxide abatement processes are summarized in Table 4-3. The details by sector are given in Appendix Tables 4A-1 through 4A-3.

Under Alternative 1, there would be negligible changes in the capital coefficients of the chemicals and primary metals sectors, and the changes for utilities would be small. Changes in operating coefficients would also be small except for utilities where there would be an increase of almost 15 percent.

There would be a moderate increase in capital coefficients under Alternative

Table 4-2
Direct Capital and Operating Costs of Sulfur Dioxide Abatement

	Capital Costs (Thousands of Dollars)			Annual Operating Costs (Thousands of Dollars)		
	(1) Year of Estimate	(2) Amount in Current Dollars	(3) Amount in 1975 Dollars	(4) Year of Estimate	(5) Amount in Current Dollars	(6) Amount in 1975 Dollars
Alternative 1						
• Chemicals	1973	$ 230	$ 273	1973	$ 580	$ 629
◦ Primary metals	1973	1,655	1,866	1973	3,465	3,760
• Electric utilities	1970	33,000	40,464	1970	11,619	13,069
Total			$ 42,603			$ 17,458
Alternative 2						
• Electric utilities	1970	$104,120	$127,369	1970	$ 32,240	$ 36,263
Total			$129,508			$ 40,652
Alternative 3						
• Electric utilities	1972	$592,660	$664,287	1972	$100,960	$109,545
Total			$666,426			$113,934

Table 4-3
Changes in Coefficients Due to Sulfur Dioxide Abatement

Expansion Capital Coefficients	Original Coefficients	New Coefficients	Difference	Percent Change
17. Chemicals	.749271	.749541	.000270	.04%
21. Primary metals	.688697	.691801	.003104	.45
46. Electrical systems:				
Alternative 1	1.751280	1.823709	.072429	4.13
Alternative 2	1.751280	1.979252	.227972	13.02
Alternative 3	1.751280	2.787364	1.036084	59.16%
Replacement Capital Coefficients				
17. Chemicals	.036380	.036394	.000014	.04%
21. Primary metals	.038410	.038612	.000202	.52
46. Electrical systems:				
Alternative 1	.053524	.055499	.001975	3.69
Alternative 2	.053524	.059736	.006212	11.61
Alternative 3	.053524	.094000	.040476	75.62%
Operating Coefficients				
17. Chemicals	.278103	.278262	.000159	.06%
21. Primary metals	.097739	.099451	.001712	1.75
46. Electrical systems:				
Alternative 1	.226387	.260129	.033742	14.90
Alternative 2	.226387	.268718	.042332	18.70
Alternative 3	.226387	.322142	.045755	42.30%

2 and also a modest rise in the operating coefficients. But there would be a large jump under Alternative 3 in both capital and operating coefficients.

Changes in total household, import, and "all other" coefficients are given in Table 4-4. The only significant changes are those in the utility sector. The change in household coefficients under Alternative 1 is quite small (2 percent), but there is a change of 18 percent as we move to Alternative 3. The changes in utility import coefficients are negative, except for Alternative 3, where there would be a rise of 12 percent.

The bottom part of Table 4-4 shows changes in total "all other" coefficients. These are negligible, except for Alternative 3, and here electric utilities would experience an increase of 5.8 percent. The sectoral details of the changes summarized in Table 4-4 are given in Appendix Table 4A-3.

We turn now to the final stage of our analysis of the cost of sulfur dioxide abatement. This follows the same format as that of Chapter 3, where the direct and indirect costs of particulate abatement were considered. The only difference is that in the sulfur dioxide case we must consider the cost of three abatement alternatives (see Table 4-5).

Table 4-4
Changes in Total Household, Import, and "All Other Final Payments" Coefficients Due to Sulfur Dioxide Abatement

Household Coefficients	Original Coefficients	Net Coefficients	Difference	Percent Change
17. Chemicals	.156220	.156221	.000001	0%
21. Primary metals	.234490	.234678	.000188	.08
46. Electrical systems:				
Alternative 1	.206190	.210645	.004455	2.16
Alternative 2	.206190	.217243	.011053	5.36
Alternative 3	.206190	.244078	.037888	18.37%
Import Coefficients				
17. Chemicals	.372450	.372451	.000001	*%
21. Primary metals	.567890	.569048	.001158	.20
46. Electrical systems:				
Alternative 1	.152810	.151243	–.001567	–1.02
Alternative 2	.152810	.147921	–.004889	–3.20
Alternative 3	.152810	.171872	.019062	12.47%
"All Other Final Payments" Coefficients				
17. Chemicals	.193200	.193207	.000007	*%
21. Primary metals	.099880	.100104	.000224	.22
46. Electrical systems:				
Alternative 1	.384620	.384620	0	0
Alternative 2	.384620	.384620	0	0
Alternative 3	.384620	.406946	.022326	5.80%

*Less than .01 percent.

When all the direct and indirect effects are taken into account, the total cost of abatement in West Virginia under Alternative 1 would be $51.3 million, and this represents an increase of only 0.3 percent in total gross output. A more significant measure is the increase in interindustry transactions required to produce a constant bill of goods for delivery to final demand. Even here, however, the increase is small in relative terms, only 0.5 percent. It is important to stress that while the strategy assumed under Alternative 1 would permit the chemicals and primary metals sectors to meet primary, and possibly secondary, standards, this would not be true of electric utilities if they limited their abatement activities to physical coal-cleaning. Thus, the cost of the remaining alternatives is more significant than the one discussed above.

The total cost of Alternative 2 is summarized in Table 4-6. When all direct and indirect costs are taken into account, the sulfur dioxide abatement bill

Table 4-5
Direct and Indirect Costs of Sulfur Dioxide Abatement under Alternative 1 (Thousands of 1975 Dollars)

		Original TGO minus Final Demand	New TGO minus Final Demand	Absolute Change	Per Cent Change
1.	Agriculture	$ 80877	$ 80893	$ 14	.02%
2.	Coal mines (undgd.)	176735	175406	-1329	-.75
3.	Coal (strip & auger)	71275	72446	1171	1.64
4.	Petroleum & natural gas	130462	130952	490	.38
5.	All other mining	84748	85439	691	.82
6.	Building contractors	508428	517361	8933	1.76
7.	Non-building contractors	165499	165978	479	.29
8.	Special contractors	337323	340190	2867	.85
9.	Food products (n.e.c.)	33214	33219	5	.02
10.	Dairies	16157	16159	2	.01
11.	Bakeries	9806	9807	1	.01
12.	Beverages	5708	5709	1	.02
13.	Apparel & accessories	6083	6085	2	.03
14.	Logging & sawmills	60817	60971	154	.25
15.	Furn. & wood fabrication	25535	25654	119	.47
16.	Printing & publishing	163432	163503	71	.04
17.	Chemicals	427069	427429	360	.08
18.	Petroleum products	98819	105800	6981	7.06
19.	Glass	28436	28445	9	.03
20.	Stone & clay products	102767	103868	1101	1.07
21.	Primary metals	179737	180461	724	.40
22.	Fabricated metals	103024	103981	957	.93
23.	Machinery (except electric)	138844	139174	330	.24
24.	Electric machinery	35244	35450	206	.58
25.	Transportation equipment	6224	6234	10	.16
26.	Instruments & products	$ 8518	$ 8519	$ 1	.01%

Table 4-5 (cont.)

		Original TGO minus Final Demand	New TGO minus Final Demand	Absolute Change	Per Cent Change
27.	All other manufacturing	$ 106525	$ 106771	$ 246	.23%
28.	Restaurants & bars	13632	13664	32	.23
29.	Wholesale trade	230900	231142	242	.10
30.	Retail food stores	2032	2033	1	.05
31.	Auto service stations	28801	28837	36	.12
32.	All other retail	125309	125681	372	.30
33.	Banking	71641	72102	461	.64
34.	Other finance	15801	15824	23	.14
35.	Insurance agents	163923	164085	162	.10
36.	Real estate	63787	63916	129	.20
37.	All other FIRE	65723	65729	6	.01
38.	Hotels & lodgings	14838	14378	-460	-3.10
39.	Medical & legal services	105899	105987	88	.08
40.	Educational services	8796	8796	0	0
41.	All other services	262326	262538	212	.08
42.	Railroads	80776	80613	-163	-.20
43.	Trucking & warehousing	164433	164403	-30	-.02
44.	All other transportation	76758	76820	62	.08
45.	Communications	121848	122007	159	.13
46.	Electrical systems	207715	208825	1110	.53
47.	Gas systems	159514	159704	190	.12
48.	Water & sanitary services	38892	39169	277	.71
	Total interindustry transactions	5140170	5162157	27960	.54
	Households	6017364	6023666	6302	.10
	Out-of-state (imports)	6079203	6096122	16919	.28
	All other final payments	3059643	3059727	84	ϕ
	Total final payments	15156210	15179575	23305	.15
	Total change in TGO	$20296380	$20331632	$ 51265	.25%

$^{\phi}$Less than .01 of one per cent.

Table 4-6
Direct and Indirect Costs of Sulfur Dioxide Abatement under Alternative 2 (Thousands of 1975 Dollars)

		Original TGO minus Final Demand	New TGO minus Final Demand	Absolute Change	Per Cent Change
1.	Agriculture	$ 80879	$ 80928	$ 49	.06%
2.	Coal mines (undgd.)	176735	178361	1626	.92
3.	Coal (strip & auger)	71275	74396	3121	4.38
4.	Petroleum & natural gas	130462	131054	592	.45
5.	All other mining	84748	85921	1173	1.38
6.	Building contractors	508428	535662	27234	5.36
7.	Non-building contractors	165499	166716	1217	.74
8.	Special contractors	337323	345889	8566	2.54
9.	Food products (n.e.c.)	33214	33233	19	.06
10.	Dairies	16157	16163	6	.04
11.	Bakeries	9806	9810	4	.04
12.	Beverages	5708	5710	2	.04
13.	Apparel & accessories	6083	6087	4	.06
14.	Logging & sawmills	60817	61305	488	.80
15.	Furn. & wood fabrication	25535	25911	376	1.47
16.	Printing & publishing	163432	163595	163	.10
17.	Chemicals	427069	427571	502	.12
18.	Petroleum products	98819	107040	8221	8.32
19.	Glass	28436	28464	28	.10
20.	Stone & clay products	102767	105987	3220	3.13
21.	Primary metals	179737	181840	2103	1.17
22.	Fabricated metals	103024	105618	2594	2.52
23.	Machinery (except electric)	138844	139664	820	.59
24.	Electric machinery	35244	35514	270	.77
25.	Transportation equipment	6224	6239	15	.24
26.	Instruments & products	$ 8518	$ 8523	$ 5	.06%

Table 4-6 (cont.)

		Original TGO minus Final Demand	New TGO minus Final Demand	Absolute Change	Per Cent Change
27.	All other manufacturing	$ 106525	$ 107262	$ 737	.69%
28.	Restaurants & bars	13632	13722	90	.66
29.	Wholesale trade	230900	231562	662	.29
30.	Retail food stores	2032	2033	1	.05
31.	Auto service stations	28801	28892	91	.32
32.	All other retail	125309	126178	869	.69
33.	Banking	71641	72839	1198	1.67
34.	Other finance	15801	15852	51	.32
35.	Insurance agents	163923	164445	522	.32
36.	Real estate	63787	64146	359	.56
37.	All other FIRE	65723	65755	32	.05
38.	Hotels & lodgings	14383	14370	-13	-.09
39.	Medical & legal services	105899	106094	195	.18
40.	Educational services	8796	8799	3	.03
41.	All other services	262326	262882	556	.21
42.	Railroads	80776	80358	-418	-.52
43.	Trucking & warehousing	164433	164191	-242	-.15
44.	All other transportation	76758	76791	33	.04
45.	Communications	121848	122068	220	.18
46.	Electrical systems	107715	210715	3000	1.44
47.	Gas systems	159514	159867	353	.22
48.	Water & sanitary services	38892	39730	838	2.15
	Total interindustry transactions	5140170	5205752	71555	1.39
	Households	6017364	6034722	17358	.29
	Out-of-state (imports)	6079203	6131278	52075	.86
	All other final payments	3059643	3059727	84	ϕ
	Total final payments	15156210	15225727	69517	.46
	Total change in TGO	$20296380	$20361962	$ 141072	.70%

$^{\phi}$Less than .01 of one per cent.

would come to $141 million, or a change of 0.7 percent in total gross output. The cost of interindustry transactions would increase by more than $71 million, or about 1.4 percent.

Finally, if both primary and secondary air quality standards in West Virginia are to be met, Alternative 3 is the only realistic one to consider.[10] This alternative would add more than $592 million to the *total* cost of producing a constant volume of goods and services for final sale and would represent an increase in total gross output of almost 3 percent. The cost of interindustry transactions would rise by more than $380 million, or 7.4 percent (see Table 4-7).

The total cost of sulfur dioxide abatement under Alternative 3 is substantially larger than that for particulate abatement (see Table 3-8). As noted earlier in this report, however, the cost of particulate and sulfur dioxide abatement have been separated only for analytical reasons. To obtain estimates of the direct and indirect cost of meeting both particulate and sulfur dioxide standards, it is necessary to combine the separate sets of estimates. The results are given in Chapter 5.

Table 4-7

Direct and Indirect Costs of Sulfur Dioxide Abatement under Alternative 3 (Thousands of 1975 Dollars)

	Original TGO minus Final Demand	New TGO minus Final Demand	Absolute Change	Per Cent Change
1. Agriculture	$ 80879	$ 81154	$ 275	.34%
2. Coal mines (undgd.)	176735	176831	96	.05
3. Coal (strip & auger)	71275	71927	652	.91
4. Petroleum & natural gas	130462	131543	1081	.83
5. All other mining	84748	89110	4362	5.15
6. Building contractors	508428	633262	124834	24.55
7. Non-building contractors	165499	197668	32169	19.44
8. Special contractors	337323	377514	40191	11.91
9. Food products (n.e.c.)	33214	33330	116	.35
10. Dairies	16157	16206	49	.30
11. Bakeries	9806	9836	30	.30
12. Beverages	5708	5721	13	.23
13. Apparel & accessories	6083	6116	33	.54
14. Logging & sawmills	60817	63140	2323	3.82
15. Furn. & wood fabrication	25535	27578	2043	8.00
16. Printing & publishing	163432	166644	3212	1.96
17. Chemicals	427069	429913	2844	.66
18. Petroleum products	98819	110111	11292	11.43
19. Glass	28436	28572	136	.48
20. Stone & clay products	102767	121744	18977	18.47
21. Primary metals	179737	192054	12317	6.85
22. Fabricated metals	103027	112964	9937	9.64
23. Machinery (except electric)	138844	158187	19343	13.93
24. Electric machinery	35244	45671	10427	29.58
25. Transportation equipment	6224	6310	86	1.38
26. Instruments & products	$ 8518	$ 8709	$ 191	2.24%

Table 4-7 (cont.)

		Original TGO minus Final Demand	New TGO minus Final Demand	Absolute Change	Per Cent Change
27.	All other manufacturing	$ 106525	$ 110158	$ 3633	3.41%
28.	Restaurants & bars	13632	14341	709	5.20
29.	Wholesale trade	230900	231283	5383	2.33
30.	Retail food stores	2032	2039	7	.34
31.	Auto service stations	28801	29342	541	1.88
32.	All other retail	125309	128739	3430	2.74
33.	Banking	71641	88879	17238	24.05
34.	Other finance	15801	16056	255	1.61
35.	Insurance agents	163923	170749	6826	4.16
36.	Real estate	63787	71502	7715	12.09
37.	All other FIRE	65723	66724	1001	1.52
38.	Hotels & lodgings	14383	14584	201	1.40
39.	Medical & legal services	105899	107449	1550	1.46
40.	Educational services	8796	8878	82	.93
41.	All other services	262326	267946	5620	2.14
42.	Railroads	80776	81474	698	.85
43.	Trucking & warehousing	164433	174162	9729	5.92
44.	All other transportation	76758	77859	1101	1.43
45.	Communications	121848	123753	1905	1.56
46.	Electrical systems	207715	217025	9310	4.48
47.	Gas systems	159514	161929	2415	1.51
48.	Water & sanitary services	38892	42838	3946	10.15
	Total interindustry transactions	5140170	5514524	380327	7.40
	Households	6017364	6047859	30495	.51
	Out-of-state (imports)	6079203	6742534	163331	2.69
	All other final payments	3059643	3077657	18014	.59
	Total final payments	15156210	15368050	211840	1.40
	Total change in TGO	$20296380	$20882574	$ 592167	2.92%

Appendix 4A

Table 4A-1
Changes in Expansion Capital Coefficients Due to the Abatement of Sulfur Dioxide Emissions

	Chemicals (17)			Primary metals (21)		
	Original b_{ij}	New b_{ij}	Difference	Original b_{ij}	New b_{ij}	Difference
6. Building contractors	.267804	.267901	.000097	.132120	.133229	.001109
7. Non-building contractors	.138557	.141669	.000050	.051069	.051643	.000574
8. Special contractors				.057900	.057900	0
15. Furn. & wood fabrication				.001987	.001987	0
20. Stone & clay products	.012753	.012758	.000005	.032448	.032501	.000053
21. Primary metals	.025286	.025295	.000009	.036235	.036340	.000105
22. Fabricated metals				.026049	.026049	0
23. Machinery (except electric)	.103712	.103749	.000037	.081213	.081643	.000430
24. Electric machinery	.016700	.016706	.000006	0	.000069	.000069
25. Transportation equipment	.000404	.000404	0	0	.000002	.000002
27. All other manufacturing				.004519	.004519	0
29. Wholesale trade	.001540	.001541	.000001	.007837	.007843	.000006
32. All other retail	.006493	.006495	.000002	.000866	.000893	.000027
50. Out-of-state (imports)	.176022	.176085	.000063	.256454	.257183	.000729

Table 4A-1 (cont.)

	Electrical systems (46)						
	Alternative 1[a]			Alternative 2[b]		Alternative 3[c]	
	Original b_{ij}	New b_{ij}	Difference	New b_{ij}	Difference	New b_{ij}	Difference
6. Building contractors	.472151	.511204	.039053	.595071	.122920	1.016747	.544596
7. Non-building contractors	.059093	.059093	0	.059093	0	.188241	.129148
8. Special contractors	.044021	.044021	0	.044021	0	.041795	-.002226
20. Stone & clay products	.013883	.015112	.001229			.026775	.012892
21. Primary metals	.015275	.017712	.002437	.017753	.003870	.049675	.034400
22. Fabricated metals				.022946	.007671		
23. Machinery (except electric)	.045257	.045257	0			.124246	.078989
24. Electric machinery	.003631	.003631	0	.045257	0	.061948	.058317
25. Transportation equipment				.003631	0	.000019	.000019
26. Instruments & products	.001901	.001901	0			.002945	.001044
27. All other manufacturing				.001901	0		
29. Wholesale trade	.020623	.020775	.000152			.021956	.001333
30. Retail food stores				.021101	.000478		
32. All other retail	.000438	.001067	.000629			.000795	.000357
33. Banking				.002418	.001980		
50. Out-of-state (imports)	1.075007	1.103936	.028929	1.166060	.091053	1.252222	.177215

a/ Physical or "deep" coal cleaning in all three sectors.

b/ Physical coal cleaning in chemicals and primary metals, and the Myers-TRW method in electric utilities.

c/ Physical coal cleaning in chemicals and primary metals, and a combination of Myers-TRW and SO_2 scrubbing in electric utilities.

Table 4A-2
Changes in Replacement Capital Coefficients Due to Sulfur Dioxide Abatement

	Chemicals (17)			Primary metals (21)		
	Original d_{ij}	New d_{ij}	Difference	Original d_{ij}	New d_{ij}	Difference
6. Building contractors	.005950	.005952	.000002	.002940	.002965	.000025
7. Non-building contractors	.003080	.003081	.000001	.001130	.001143	.000013
8. Special contractors				.003620	.003620	0
15. Furn. and wood fabrication				.000120	.000120	0
20. Stone & clay products	.000280	.000280	0	.000720	.000721	.000001
21. Primary metals	.000560	.000560	0	.000810	.000812	.000002
22. Fabricated metals				.001630	.001630	0
23. Machinery (except electric)	.009430	.009434	.000004	.005080	.005107	.000027
24. Electric machinery	.001520	.001521	.000001	0	.000069	.000069
25. Transportation equipment	.000040	.000040	0	0	.000002	.000002
27. All other manufacturing				.000280	.000280	0
29. Wholesale trade	.000140	.000140	0	.000490	.000490	0
32. All other retail	.000590	.000590	0	.000050	.000051	.000001
50. Out-of-state (imports)	.014790	.014796	.000006	.021540	.021602	.000062

Table 4A-2 (cont.)

	Electrical systems (46)						
		Alternative 1[a/]		Alternative 2[b/]		Alternative 3[c/]	
	Original	New	Difference	New	Difference	New	Difference
6. Building contractors	.010490	.011358	.140950	.013221	.002731	.022996	.012506
7. Non-building contractors	.001310	.001310	0	.001310	0	.003952	.002642
8. Special contractors	.001520	.001520	0	.001520	0	.001520	0
20. Stone & clay products	.000310	.000338	.000028	.000396	.000086	.000609	.000299
21. Primary metals	.000339	.000393	.000054	.000509	.000170	.001113	.000774
23. Machinery (except electric)	.001560	.001560	0	.001560	0	.008875	.007315
24. Electric machinery	.000130	.000130	0	.000130	0	.003181	.003051
26. Instruments & products	.000070	.000070	0	.000070	0	.000070	0
29. Wholesale trade	.000710	.000715	.000005	.000727	.000017	.000710	0
32. All other retail	.000015	.000037	.000022	.000083	.000068	.000010	-.000005
50. Out-of-state (imports)	.037070	.038068	.000998	.040210	.003140	.050964	.013894

a/ Physical or "deep" coal cleaning in all three sectors.

b/ Physical coal cleaning in chemicals and primary metals, and the Myers-TRW method in electric utilities.

c/ Physical coal cleaning in chemicals and primary metals, and a combination of Myers-TRW and SO_2 scrubbing in electric utilities.

Table 4A-3
Changes in Technical (Input) Coefficients Due to Sulfur Dioxide Abatement

	Chemicals (17)			Primary metals (21)		
	Original a_{ij}	New a_{ij}	Difference	Original a_{ij}	New a_{ij}	Difference
1. Agriculture	.000010	.000010	0	.001781	.001778	-.000003
2. Coal mines (undgd.)	.010372	.010166	-.000226	.019301	.018572	-.000729
4. Petroleum & natural gas	.007533	.007532	-.000001	.000197	.000198	.000001
5. All other mining	.004321	.004320	-.000001	.004606	.004597	-.000009
6. Building contractors	.000846	.000846	0	.000022	.000022	0
7. Non-building contractors	.001245	.001245	0	.000044	.000044	0
8. Special contractors	.005750	.005749	-.000001	.001883	.001880	-.000003
9. Food products (n.e.c.)	.000402	.000402	0			
13. Apparel & accessories	.000073	.000073	0	.000007	.000007	0
14. Logging & sawmills				.000007	.000007	0
15. Furn. & wood fabrication	.000108	.000108	0			
16. Printing & publishing	.001200	.001200	0	.000128	.000128	0
17. Chemicals	.146151	.146125	-.000026	.003803	.003818	.000015
18. Petroleum products	.001043	.001457	.000414	.000102	.002635	.002533
19. Glass	.000622	.000622	0	.000279	.000278	-.000001
20. Stone & clay products	.001132	.001132	0			
21. Primary metals	.008294	.008293	-.000001	.003976	.003970	-.000006
22. Fabricated metals	.000677	.000677	0	.001527	.001525	-.000002
23. Machinery (except electric)				.000457	.000456	-.000001
24. Electric machinery	.000945	.000945	0	.000380	.000401	.000021
25. Transportation equipment				.000068	.000068	0
26. Instruments & products	.001706	.001705	-.000001			
27. All other manufacturing	.007466	.007465	-.000001	.001022	.001022	0

Table 4A-3 (cont.)

	Chemicals (17)			Primary metals (21)		
	Original a_{ij}	New a_{ij}	Difference	Original a_{ij}	New a_{ij}	Difference
28. Restaurants & bars	.000327	.000327	0	.000043	.000043	0
29. Wholesale trade	.013666	.013663	-.000003	.004786	.004779	-.000007
30. Retail food stores	.000104	.000104	0			
31. Auto service stations	.000038	.000038	0	.000003	.000003	0
32. All other retail	.001527	.001527	0	.002533	.002528	-.000005
33. Banking	.000402	.000402	0	.000980	.000978	-.000002
34. Other finance				.000165	.000164	-.000001
35. Insurance agents	.003106	.003105	-.000001	.000760	.000759	-.000001
37. All other FIRE	.001161	.001161	0	.000633	.000632	-.000001
38. Hotels & lodgings	.000731	.000731	0	.000189	.000188	-.000001
39. Medical & legal services	.001407	.001406	-.000001	.001164	.001162	-.000002
40. Educational services	.000156	.000156	0	.000398	.000397	-.000001
41. All other services	.002722	.002721	-.000001	.007939	.007924	-.000015
42. Railroads	.005646	.005645	-.000001	.008492	.008476	-.000016
43. Trucking & warehousing	.013383	.013380	-.000003	.008236	.008222	-.000014
44. All other transportation	.003504	.003503	-.000001	.009152	.009135	-.000017
45. Communications	.003701	.003700	-.000001	.002563	.002559	-.000004
46. Electrical systems	.009747	.009745	-.000002	.007231	.007218	-.000013
47. Gas systems	.015763	.015760	-.000003	.002317	.002314	-.000003
48. Water & sanitary services	.001116	.001116	0	.000565	.000564	-.000001
49. Households	.156220	.156221	.000001	.234490	.234678	.000188
50. Out-of-state (imports)	.372450	.372451	.000001	.567890	.569048	.001158
51. All other final payments	.193200	.193207	.000007	.099880	.100104	.000224

Table 4A-3 (cont.)

	Electrical systems (46)						
		Alternative 1[a/]		Alternative 2[b/]		Alternative 3[c/]	
	Original a_{ij}	New a_{ij}	Difference	New a_{ij}	Difference	New a_{ij}	Difference
2. Coal mines (undgd.)	.089369	.089521	.000152	.092559	.003190	.089468	.000099
3. Coal (strip & auger)	.003872	.035013	.031141	.036837	.002965	.033915	.000043
8. Special contractors	.000011	.000011	0	.000010	-.000001	.000010	-.000001
15. Furn. & wood fabrication	.000199	.000196	-.000003	.000189	-.000010	.000198	-.000001
16. Printing & publishing	.003954	.003890	-.000064	.003768	-.000186	.004742	.000788
17. Chemicals						.000945	.000945
18. Petroleum products	.005274	.006035	.000761	.007304	.002030	.005273	-.000001
20. Stone & clay products	.002595	.002892	.000297	.003384	.000789	.011183	.008588
21. Primary metals	.000464	.000457	-.000007	.000443	-.000021	.000592	.000128
22. Fabricated metals	.006239	.006645	.000406	.007313	.001074	.007970	.001731
24. Electric machinery	.000199	.000196	-.000003	.000189	-.000010	.000298	.000099
27. All other manufacturing	.000007	.000007	0	.000007	0	.000007	0
28. Restaurants & bars	.000620	.000610	-.000010	.000591	-.000029	.000791	.000171
29. Wholesale trade	.003289	.003235	-.000054	.003134	-.000155	.004720	.001431
31. Auto service stations	.000083	.000082	-.000001	.000079	-.000004	.000083	0
32. All other retail	.000578	.000569	-.000009	.000551	-.000027	.000750	.000172
33. Banking	.000051	.000439	.000388	.001097	.001046	.019387	.019336
35. Insurance agents	.000902	.000881	-.000021	.000859	-.000043	.005338	.004436
36. Real estate	.001153	.001134	-.000019	.001099	-.000054	.008452	.007299
37. All other FIRE	.001905	.001874	-.000031	.001815	-.000090	.001904	-.000001
38. Hotels & lodgings	.001153	.001134	-.000019	.001099	-.000054	.001153	0
39. Medical & legal services	.001272	.001251	-.000021	.001212	-.000050	.001272	0

Table 4A-3 (cont.)

	Electrical systems (46)						
		Alternative 1[a]		Alternative 2[b]		Alternative 3[c]	
	Original a_{ij}	New a_{ij}	Difference	New a_{ij}	Difference	New a_{ij}	Difference
40. Educational services	.000007	.000007	0	.000007	0	.000007	0
41. All other services	.003383	.003328	-.000055	.003224	-.000159	.004585	.001202
42. Railroads	.013507	.013286	-.000221	.012872	-.000635	.013507	0
43. Trucking & warehousing	.020820	.020479	-.000341	.019840	-.000980	.028791	.007971
44. All other transportation	.003813	.003757	-.000056	.003634	-.000179	.003813	0
45. Communications	.004869	.004787	-.000082	.004640	-.000229	.004868	-.000001
46. Electrical systems	.055628	.057261	.001633	.059847	.004219	.066950	.011322
47. Gas systems	.001101	.001083	-.000018	.001049	-.000052	.001100	-.000001
48. Water & sanitary services	.000070	.000069	-.000001	.000067	-.000003	.000070	0
49. Households	.206190	.210645	.004455	.217243	.011053	.244078	.037880
50. Out-of-state (imports)	.152810	.151243	-.001567	.147921	-.004889	.171872	.019062
51. All other final payments	.384620	.377983	-.006637	.366087	-.018533	.406946	.022326

a/ Physical or "deep" coal cleaning in all three sectors.

b/ Physical coal cleaning in chemicals and primary metals, and the Myers-TRW method in electric utilities.

c/ Physical coal cleaning in chemicals and primary metals, and a combination of Myers-TRW and SO_2 scrubbing in electric utilities.

Table 4A-4
Minimal Direct Capital and Operating Costs of Sulfur Dioxide Abatement

	Capital Costs (Thousands of Dollars)			Direct Annual Operating Costs (Thousands of Dollars)		
	(1) Year of Estimate	(2) Amount in Current Dollars	(3) Amount in 1975 Dollars	(4) Year of Estimate	(5) Amount in Current Dollars	(6) Amount in 1975 Dollars
Alternative 1						
Chemicals	1973	$ 230	$ 273	1973	$ 580	$ 629
Primary metals	1973	1,655	1,866	1973	3,465	3,760
Electric utilities	1970	33,000	40,464	1970	11,619	13,069
Total			$ 42,603			$ 17,458
Alternative 2						
Electric utilities	1970	$104,120	$127,369	1970	$ 32,240	$ 36,263
Total			$129,508			$ 40,652
Alternative 3						
Electric utilities	1972	$305,040	$341,645	1972	$100,960	$109,545
Total			$343,784			$113,934

Note: This is a reproduction of Table 4-2 in the text except for row 5, electric utilities under Alternative 3. Here we have estimated the capital cost at $48 per kw of capacity. This is the minimal capital cost estimate based on the assumption that *all* of the investment will be in *new* capacity. The higher figure in the text ($96 per kw of capacity) takes into account the costs of retrofitting in existing plants as well as investment in new capacity.

Table 4A-5

Direct and Indirect Costs of Sulfur Dioxide Abatement under the Minimal Direct Capital Cost Assumptions of Table 4A-4

	Original TGO minus Final Demand	New TGO minus Final Demand	Absolute Change	Per Cent Change
1. Agriculture	$ 80879	$ 81035	$ 156	.19%
2. Coal mines (undgd.)	176735	176442	-293	-.16
3. Coal (strip & auger)	71275	71837	562	.79
4. Petroleum & natural gas	130462	131274	812	.62
5. All other mining	84748	87379	2631	3.10
6. Building contractors	508428	572102	63674	12.52
7. Non-building contractors	165499	182532	17033	10.29
8. Special contractors	337323	358021	20698	6.14
9. Food products (n.e.c.)	33214	33284	70	.21
10. Dairies	16157	16189	32	.20
11. Bakeries	9806	9826	20	.20
12. Beverages	5708	5717	9	.16
13. Apparel & accessories	6083	6104	21	.34
14. Logging & sawmills	60817	62031	1214	2.00
15. Furn. & wood fabrication	25535	26610	1075	4.21
16. Printing & publishing	163432	165824	2392	1.46
17. Chemicals	427069	429135	2066	.48
18. Petroleum products	98819	107886	9067	9.18
19. Glass	28436	28512	76	.27
20. Stone & clay products	102767	115683	12916	12.57
21. Primary metals	179737	186287	6550	3.64
22. Fabricated metals	103024	108963	5939	5.76
23. Machinery (except electric)	$ 138844	$ 149148	$ 10304	7.42%

Table 4A-5 (cont.)

	Original TGO minus Final Demand	New TGO minus Final Demand	Absolute Change	Per Cent Change
24. Electric machinery	$ 35244	$ 40815	$ 5571	15.81%
25. Transportation equipment	6224	6280	56	.90
26. Instruments & products	8518	8637	119	1.40
27. All other manufacturing	106525	108451	1926	1.81
28. Restaurants & bars	13632	14101	469	3.44
29. Wholesale trade	230900	234641	3741	1.62
30. Retail food stores	2032	2037	5	.25
31. Auto service stations	28801	29131	330	1.14
32. All other retail	125309	127686	2377	1.90
33. Banking	71641	80586	8945	12.48
34. Other finance	15801	15949	148	.94
35. Insurance agents	163923	169321	5398	3.29
36. Real estate	63787	70824	7037	11.03
37. All other FIRE	65723	66401	678	1.03
38. Hotels & lodgings	14383	14506	123	.85
39. Medical & legal services	105899	106871	972	.92
40. Educational services	8796	8840	44	.50
41. All other services	262326	265995	3669	1.40
42. Railroads	80776	81266	490	.61
43. Trucking & warehousing	164433	172908	8475	5.15
44. All other transportation	76758	77441	683	.89
45. Communications	121848	123099	1251	1.03
46. Electrical systems	207715	216112	8397	4.04
47. Gas systems	159514	161046	1532	.96
48. Water & sanitary services	38892	40932	2040	5.24
Total interindustry transactions	5134197	5355697	221500	4.31
Households	6017364	6047859	30495	.51
Out-of-state (imports)	6079203	6163785	84582	1.39
All other final payments	3059643	3077657	18014	.59
Total final payments	15146698	15289301	142603	.94
Total change in TGO	$ 20280895	$ 20644998	$ 364103	1.80%

5 The Total Cost of Meeting Clean Air Standards

This short chapter adds nothing new in terms of data or findings. It brings together the separate estimates of the cost of meeting particulate and sulfur dioxide standards discussed separately in Chapters 3 and 4.

Since three alternative processes for dealing with sulfur emissions were given in the previous chapter, it is necessary for us to present all total cost estimates in terms of the three alternatives.

The Direct Capital and Operating Costs of Particulate and Sulfur Abatement

The combined costs of complying with particulate and sulfur emission standards are given in Table 5-1. The first column shows the cost of installing abatement plant and equipment, and the second shows annual operating costs by sector for each of the three alternatives. Both are expressed in terms of 1975 estimated prices.

The total capital investment required under Alternative 1 would be slightly more than \$352 million. Earlier, we had estimated that the West Virginia economy will produce about \$13.6 billion worth of goods and services in 1975 (when expressed in 1975 dollars). Thus, the total capital cost under Alternative 1 would be about 2.5 percent of the estimated Gross State Product. Annual operating costs of \$55 million would be about 0.4 percent of GSP. In relative terms, the aggregate cost of Alternative 1 would not be particularly high. But while this alternative would take care of most of West Virgnia's particulate problems, it would not permit public utilities in the state to meet the primary standards established by the WVAPCC, much less the more stringent secondary standards.

Alternative 2 would go further toward this goal. But a successful chemical coal-cleaning process, such as the Myers-TRW method discussed in Chapter 4, would add substantially to the total cost of pollution abatement in West Virginia. Capital expenditures would go up to about \$439 million, and this would amount to about 3.2 percent of estimated GSP. Annual operating costs would increase to \$78 million, or about 0.6 percent of GSP. But Alternative 2 would not completely accomplish the goal of meeting established clean air standards.

We turn, finally, to Alternative 3, where a combination of chemical coal-

Table 5-1
Direct Capital and Operating Costs of Particulate and Sulfur Dioxide Abatement (Thousands of 1975 Dollars)

	Alternative 1[a]		Alternative 2[a]		Alternative 3[a]	
	Capital Cost	Operating Cost	Capital Cost	Operating Cost	Capital Cost	Operating Cost
2. Coal mines (undgd.)	$ 2,409	$ 532	$ 2,409	$ 532	$ 2,409	$ 532
9. Food products (n.e.c.)	245	49	245	49	245	49
10. Dairies	244	45	244	45	244	45
11. Bakeries	183	34	183	34	183	34
12. Beverages	244	45	244	45	244	45
17. Chemicals	46,997	14,466	46,997	14,466	46,997	14,466
18. Petroleum products	112	217	112	217	112	217
19. Glass	560	1,085	560	1,085	560	1,085
20. Stone & clay products	3,597	697	3,597	697	3,597	697
21. Primary metals	89,824	20,474	89,824	20,474	89,824	20,474
22. Fabricated metals	392	760	392	760	392	760
23. Machinery (except electric)	2,112	332	2,112	332	2,112	332
24. Electric machinery	1,043	164	1,043	164	1,043	164
25. Transportation equipment	2,394	377	2,394	377	2,394	377
26. Instruments & products	197	310	197	310	197	310
46. Electrical systems	201,473	15,658	288,378	38,852	825,296	112,134
Total	$352,026	$55,245	$438,931	$78,439	$975,849	$151,721

a/ The detailed assumptions behind these alternatives are given in Chapter 4.

cleaning and sulfur oxide flue-gas scrubbing would satisfy the primary and secondary standards established by Regulation X. Under this alternative, the capital cost jumps to almost $976 million, and this would amount to about 7.2 percent of estimated GSP. Annual direct operating costs of nearly $152 million would account for 1.1 percent of GSP under Alternative 3. It is evident that if we are serious about reaching the goal of clean air in West Virginia, we must be prepared to pay a substantial bill. And the cost estimates given above relate only to pollution generated by stationary sources; we have not taken into account the large contribution that automobiles make to total air pollution, nor have we considered the cost of cleaning up air pollution generated by other moving sources. Furthermore, Table 5-1. deals only with the *direct* costs of particulate abatement. It is necessary to examine the direct plus indirect costs to obtain a clearer perspective of what industrial compliance with West Virginia clean air standards will cost.

The Direct and Indirect Costs of Clean Air

The total cost of controlling particulate emissions and the cost of achieving partial control of sulfur emissions under Alternative 1 are given in Table 5-2. These cost increases are expressed in terms of changes in output, and the figures in Column 4 of Table 5-2 should not be construed as indicators of the incidence of abatement cost. They indicate, instead, the *total* increases in interindustry transactions required to produce a constant bill of goods for delivery to final demand.

In Chapter 3, we showed that only 16 industries would be directly affected to a significant extent by particulate regulation, and Chapter 4 indicated that only 3 West Virginia sectors are considered to be major emitters of sulfur oxides. But because of interdependence on both the input and output sides, all sectors of the economy contribute indirectly to air pollution.

Some economic activities in West Virginia will experience economic gains from the installation and operation of pollution abatement equipment. This is particularly true of general building contractors and of banks and other financial institutions that will help finance the large capital outlays that will be required. But the *total* changes in output may be considered as accurate reflections of cost changes since they are the changes in production that would be required to produce a constant vector of final demand, which we have assumed will amount to $13.6 billion in 1975. The most significant changes are those at the bottom of the table. There we see that total interindustry transactions would increase by $144 million, or 2.18 percent. Imports would go up by $163 million, or 2.7 percent. Payments to households, primarily for labor services, would increase almost $13 million, or 0.2 percent. The total increase in TGO would be $323 million, or about 1.6 percent.

Table 5-2
Direct and Indirect Costs of Particulate and Sulfur Abatement under Alternative 1

	(1) Original TGO less Constant Final Demand	(2) Cost of Particulate Abatement	(3) Cost of Sulfur Abatement	(4) Total Cost of Abatement (2) + (3)	(5) Per Cent Increase in Production Cost (4) as % of (1)
1. Agriculture	$ 80,879	$ 123	$ 14	$ 137	.17%
2. Coal mines (undgd.)	176,735	2,794	-1,329	1,465	.83
3. Coal (strip & auger)	71,275	-226	1,171	945	1.33
4. Petroleum & natural gas	130,461	196	490	686	.53
5. All other mining	84,747	736	691	1,427	1.68
6. Building contractors	508,428	44,742	8,933	53,675	10.56
7. Non-building contractors	165,499	3,710	479	4,189	2.53
8. Special contractors	337,323	17,385	2,867	20,252	6.00
9. Food products (n.e.c.)	33,214	75	5	80	.24
10. Dairies	16,157	10	2	12	.07
11. Bakeries	9,805	7	1	8	.08
12. Beverages	5,707	3	1	4	.07
13. Apparel & accessories	6,082	7	2	9	.15
14. Logging ¢ sawmills	60,816	879	154	1,033	1.70
15. Furn. & wood fabrication	25,534	833	119	952	3.73
16. Printing & publishing	163,837	537	71	608	.37
17. Chemicals	427,068	4,868	360	5,228	1.22
18. Petroleum products	99,089	696	6,981	7,677	7.75
19. Glass	28,435	4	9	13	.05
20. Stone & clay products	102,766	494	1,101	1,595	1.55
21. Primary metals	179,736	3,856	724	4,580	2.55
22. Fabricated metals	103,023	3,924	957	4,881	4.74
23. Machinery (except electric)	138,843	6,622	330	6,952	5.01
24. Electric machinery	35,244	793	206	999	2.83
25. Transportation equipment	6,223	35	10	45	.72
26. Instruments & products	8,517	-4	1	-3	-.04%

Table 5-2 (cont.)

	(1) Original TGO less Constant Final Demand	(2) Cost of Particulate Abatement	(3) Cost of Sulfur Abatement	(4) Total Cost of Abatement (2) + (3)	(5) Per Cent Increase in Production Cost (4) as % of (1)
27. All other manufacturing	$ 106,525	$ 806	$ 246	$ 1,052	.99%
28. Restaurants & bars	13,631	169	32	201	1.47
29. Wholesale trade	230,899	1,803	242	2,045	.89
30. Retail food stores	2,032	2	1	3	.15
31. Auto service stations	28,782	176	36	212	.74
32. All other retail	125,249	1,369	372	1,741	1.39
33. Banking	71,641	7,806	461	8,467	11.82
34. Other finance	15,800	75	23	98	.62
35. Insurance agents	163,922	897	162	1,059	.65
36. Real estate	63,786	506	129	635	1.00
37. All other FIRE	65,723	218	6	224	.34
38. Hotels & lodgings	14,382	36	-460	-424	-2.95
39. Medical & legal services	105,899	340	88	428	.40
40. Educational services	8,796	6	0	6	.07
41. All other services	262,326	904	212	1,116	.43
42. Railroads	80,775	-104	-163	-267	-.33
43. Trucking & warehousing	164,433	852	-30	822	.50
44. All other transportation	76,757	-109	62	- 47	-.06
45. Communications	121,848	278	159	437	.36
46. Electrical systems	207,715	4,372	1,110	5,482	2.64
47. Gas systems	159,513	1,576	190	1,766	1.11
48. Water & sanitary services	38,891	1,349	277	1,626	4.18
Total interindustry transact.	5,140,170	116,426	27,960	144,386	2.81
Households	6,017,364	6,401	6,302	12,703	.21
Out-of-state (imports)	6,079,203	146,252	16,919	163,171	2.68
All other final payments	3,059,643	2,417	84	2,501	.08
Total final payments	15,156,210	155,070	23,305	178,375	1.18
Total change in TGO	$20,296,380	$271,496	$ 51,265	$322,761	1.59%

Table 5-3 shows the same changes under Alternative 2. As indicated above, this alternative would be an improvement over physical coal-cleaning alone, but it would not permit utilities to achieve established emission standards. If Alternative 2 were adopted, interindustry transactions would increase by $88 million, or 3.7 percent. Household payments would rise almost $24 million (0.4 percent), and imports would increase by $195 million, or 3.3 percent. TGO would go up almost $413 million, or more than 2 percent.

The only alternative that would meet both primary and secondary particulate and sulfur standards under Regulation X is Alternative 3. If this alternative were adopted, there would be no further degradation of air quality from stationary sources in West Virginia. But the total cost would be substantial (see Table 5-4). The interindustry transactions required to produce a constant bill of goods under this alternative would go up by almost $496 million, or 9.7 percent. Payments to households would increase by almost $37 million (0.6 percent). And imports would rise more than $309 million, or 5.1 percent. Total gross output would rise almost $864 million, or 4.3 percent to produce a constant volume of goods and services for delivery to final demand.

There are substantial sector-to-sector variations in the three alternative methods we have discussed. The pattern is one of steadily increasing costs (measured by changes in sectoral output) as we move from Alternative 1 to Alternative 3. The total economic activity generated by a combination of coal-cleaning and effective flue-gas scrubbing would increase output in construction, as well as in some of the manufacturing sectors. As indicated earlier, there also would be major increases in financial activity within the state.

The Regional Economic Consequences of Clean Air

This study has focused entirely on the costs involved in industrial compliance with West Virginia's clean air standards. What are the economic consequences? Cost increases of the magnitude we have discussed in this report cannot be "absorbed" by industrial and commercial establishments. These increases will have to be passed on to consumers in the form of higher prices. But the cost increases are not expected to place West Virginia industry at a competitive disadvantage. Since clean air legislation is national in scope, manufacturing industry, wherever it is located, will be required to comply with established standards.

Even before the National Energy Emergency Bill of 1973 was introduced, there was evidence that both the Environmental Protection Agency and state commissions were prepared to make exceptions, and grant variances, when individual companies were able to present convincing evidence that they needed more time to meet the standards. This was particularly true in the case of secondary standards. If an energy act containing the major provisions of the

Table 5-3
Direct and Indirect Costs of Particulate and Sulfur Abatement under Alternative 2

	(1) Original TGO less Constant Final Demand	(2) Cost of Particulate Abatement	(3) Cost of Sulfur Abatement	(4) Total Cost of Abatement (2) + (3)	(5) Per Cent Increase in Production Cost (4) as % of (1)
1. Agriculture	$ 80,879	$ 123	$ 49	$ 172	.21%
2. Coal mines (undgd.)	176,735	2,794	1,626	4,420	2.50
3. Coal (strip & auger)	71,275	-226	3,121	2,895	4.06
4. Petroleum & natural gas	130,461	196	592	788	.60
5. All other mining	84,747	736	1,173	1,909	2.25
6. Building contractors	508,428	44,742	27,234	71,976	14.16
7. Non-building contractors	165,499	3,710	1,217	4,927	2.98
8. Special contractors	337,323	17,385	8,566	25,951	7.69
9. Food products (n.e.c.)	33,214	75	19	94	.28
10. Dairies	16,157	10	6	16	.10
11. Bakeries	9,805	7	4	11	.11
12. Beverages	5,707	3	2	5	.09
13. Apparel & accessories	6,082	7	4	11	.18
14. Logging & sawmills	60,816	879	488	1,367	2.25
15. Furn. & wood fabrication	25,534	833	376	1,209	4.73
16. Printing & publishing	163,837	537	163	700	.43
17. Chemicals	427,068	4,868	502	5,370	1.26
18. Petroleum products	99,089	696	8,221	8,917	9.00
19. Glass	28,435	4	28	32	.11
20. Stone & clay products	102,766	494	3,220	3,714	3.61
21. Primary metals	179,736	3,856	2,130	5,986	3.33
22. Fabricated metals	103,023	3,924	2,594	6,518	6.33
23. Machinery (except electric)	138,843	6,622	820	7,442	5.36
24. Electric machinery	35,244	793	270	1,066	3.02
25. Transportation equipment	6,223	35	15	50	.80
26. Instruments & products	8,517	-4	5	1	.01%

Table 5-3 (cont.)

	(1) Original TGO less Constant Final Demand	(2) Cost of Particulate Abatement	(3) Cost of Sulfur Abatement	(4) Total Cost of Abatement (2) + (3)	(5) Per Cent Increase in Production Cost (4) as % of (1)
27. All other manufacturing	$ 106,525	$ 806	$ 1,779	$ 2,585	2.43%
28. Restaurants & bars	13,631	169	90	259	1.90
29. Wholesale trade	230,899	1,803	662	2,425	1.05
30. Retail food stores	2,032	2	1	3	.15
31. Auto service stations	28,782	176	91	267	.93
32. All other retail	125,249	1,369	869	2,238	1.79
33. Banking	71,641	7,806	1,198	9,004	12.57
34. Other finance	15,800	75	51	126	.80
35. Insurance agents	163,922	897	522	1,419	.87
36. Real estate	63,786	506	359	865	1.36
37. All other FIRE	65,723	218	32	250	.38
38. Hotels & lodgings	14,382	36	-13	23	.16
39. Medical & legal services	105,899	340	195	535	.51
40. Educational services	8,796	10	3	9	.10
41. All other services	262,326	904	556	1,460	.56
42. Railroads	80,775	-104	-418	522	.65
43. Trucking & warehousing	164,433	852	-242	610	.37
44. All other transportation	76,757	-109	33	76	.10
45. Communications	121,848	278	220	498	.41
46. Electrical systems	207,715	4,372	3,000	7,372	3.55
47. Gas systems	159,513	1,576	353	1,929	1.21
48. Water & sanitary services	38,891	1,349	838	2,187	5.62
Total interindustry transact.	5,140,170	116,426	71,555	187,981	3.66
Households	6,017,364	6,401	17,358	23,759	.39
Out-of-state (imports)	6,079,203	146,252	52,075	198,327	3.26
All other final payments	3,059,643	2,417	84	2,501	.08
Total final payments	15,156,210	155,070	69,519	224,589	1.48
Total change in TGO	$20,296,380	$271,496	$141,072	$412,570	2.03%

Table 5-4
Direct and Indirect Costs of Particulate and Sulfur Abatement under Alternative 3

	(1) Original TGO less Constant Final Demand	(2) Cost of Particulate Abatement	(3) Cost of Sulfur Abatement	(4) Total Cost of Abatement (2) + (3)	(5) Per Cent Increase in Production Cost (4) as % of (1)
1. Agriculture	$ 80,879	$ 123	$ 275	$ 398	.49%
2. Coal mines (undgd.)	176,735	2,794	96	2,890	1.64
3. Coal (strip & auger)	71,275	-226	652	426	.60
4. Petroleum & natural gas	130,461	196	1,081	1,277	.98
5. All other mining	84,747	736	4,362	5,108	6.03
6. Building contractors	508,428	44,742	124,834	169,576	33.35
7. Non-building contractors	165,499	3,710	32,169	35,879	21.68
8. Special contractors	337,323	17,385	40,191	57,576	17.07
9. Food products (n.e.c.)	33,214	75	116	191	.58
10. Dairies	16,157	10	49	59	.37
11. Bakeries	9,805	7	30	37	.41
12. Beverages	5,707	3	13	16	.28
13. Apparel & accessories	6,082	7	33	40	.66
14. Logging & sawmills	60,816	879	2,323	3,202	5.27
15. Furn. & wood fabrication	25,534	833	2,043	2,876	11.26
16. Printing & publishing	163,837	537	3,212	3,749	2.29
17. Chemicals	427,068	4,868	2,844	7,712	1.81
18. Petroleum products	99,089	696	11,292	11,988	12.10
19. Glass	28,435	4	136	140	.49
20. Stone & clay products	102,766	494	18,977	19,471	18.95
21. Primary metals	179,736	3,856	12,317	16,173	9.00
22. Fabricated metals	103,023	3,924	9,937	13,861	13.45
23. Machinery (except electric)	138,843	6,622	19,343	25,965	18.70
24. Electric machinery	35,244	793	10,427	11,220	31.84
25. Transportation equipment	6,223	35	86	121	1.94
26. Instruments & products	$ 8,517	$ -4	$ 191	$ 187	2.20%

Table 5-4 (cont.)

	(1) Original TGO less Constant Final Demand	(2) Cost of Particulate Abatement	(3) Cost of Sulfur Abatement	(4) Total Cost of Abatement (2) + (3)	(5) Per Cent Increase in Production Cost (4) as % of (1)
27. All other manufacturing	$ 106,525	$ 806	$ 3,633	$ 4,439	4.17%
28. Restaurants & bars	13,631	169	709	878	6.44
29. Wholesale trade	230,899	1,803	5,383	7,186	3.11
30. Retail food stores	2,032	2	7	9	.44
31. Auto service stations	28,782	176	741	917	3.19
32. All other retail	125,249	1,369	3,430	4,799	3.83
33. Banking	71,641	7,806	17,238	25,044	34.96
34. Other finance	15,800	75	255	330	2.09
35. Insurance agents	163,922	897	6,826	7,723	4.71
36. Real estate	63,786	506	7,715	8,221	12.89
37. All other FIRE	65,723	218	1,001	1,219	1.85
38. Hotels & lodgings	14,382	36	201	237	1.65
39. Medical & legal services	105,899	340	1,550	1,890	1.78
40. Educational services	8,796	6	82	88	1.00
41. All other services	262,326	904	5,620	6,524	2.49
42. Railroads	80,775	-104	698	594	.74
43. Trucking & warehousing	164,433	852	9,729	10,581	6.43
44. All other transportation	76,757	-109	1,101	992	1.29
45. Communications	121,848	278	1,905	2,183	1.79
46. Electrical systems	207,715	4,372	9,310	13,682	6.59
47. Gas systems	159,513	1,576	2,415	3,991	2.50
48. Water & sanitary services	38,891	1,349	3,946	5,295	13.61
Total interindustry transact.	5,140,170	116,426	$380,327	496,753	9.66
Households	6,017,364	6,401	30,495	36,896	.61
Out-of-state (imports)	6,079,203	146,252	163,331	309,583	5.09
All other final payments	3,059,643	2,417	18,014	20,431	.67
Total final payments	15,156,210	155,070	211,840	366,910	2.42
Total change in TGO	$20,296,380	$271,496	$592,167	$ 863,663	4.26%

1973 bill is passed, the result could be a whole new ball game. The 1973 bill, for example, would have extended the date for compliance from July 1, 1975 to July 1, 1977. Presumably, however, industry will eventually have to take the steps discussed in this report.

One final point should be made about the cost estimates developed in this study. They relate specifically to West Virginia industrial sectors. But the *direct* cost estimates may be more widely applicable. If adjustments can be made for differences in industry-mix from region to region, the West Virginia estimates could serve as the basis for similar estimates in other regions. We would not expect, however, that the changes in interindustry transactions to produce a constant bill of goods calculated for West Virginia would be applicable elsewhere because no two states have identical economic structures.

6 Simulating Low BTU Coal Gasification

Preceding chapters have discussed the potential negative effects of the Clean Air Act of 1970 on the economic development of West Virginia. Early in the study we hypothesized that industrialists faced with major investment decisions might consider relocation of some West Virginia facilities—particularly older plants—rather than installation of expensive abatement equipment in existing facilities. We found no evidence to support this hypothesis, however. Although full compliance with the 1970 air quality standards will raise production costs in those West Virginia industries that are heavy air polluters, there will be similar cost increases throughout the nation. Thus air pollution control legislation is not expected to have *negative* effects on West Virginia's future industrial development.[1] Is it possible, however, that there could be *positive* effects? Is there any way in which federal efforts to reduce air pollution could accelerate the rate of economic growth in West Virginia? The present chapter attempts to answer these questions.

The Energy Shortage

The "energy crisis" has been widely discussed and little need be said about it here. The long-range problem that comes under this rubric has been summarized by Chauncey Starr, former Dean of the School of Engineering and Applied Sciences at the University of California at Los Angeles:

> Between now and 2001, just 30 years away, the United States will consume more energy than it has in its entire history. By 2001 the annual U.S. demand for energy in all forms is expected to double, and the annual worldwide demand will probably triple. These projected increases will tax man's ability to discover, extract, and refine fuels in the huge volumes necessary, to ship them safely, to find suitable locations for several hundred new electric-power stations in the United States (thousands worldwide) and to dispose of effluents and waste products with minimum harm to himself and his environment.[2]

Energy is available to satisfy our national needs for centuries. But one of the primary domestic sources of energy—coal—is now being used inefficiently and in ways that contribute heavily to atmospheric pollution. It has been estimated, for example, that the combustion of coal in stationary plants emitted more than 33 million tons of air pollutants in 1968.[3] At present, much of the available energy stored in coal cannot be converted to usable form without undesirable environmental side effects. One reason for this is a technological lag in the conversion of fossil fuels to "clean" fuels. This, in turn, is to a large extent the result of faulty forecasting and the virtual absence of energy planning in the past.

During the 1950s there was a major shift in demand from coal to oil, particularly in transportation. This shift in demand was one of the more important causes of growing economic distress throughout rural Appalachia during this period. There were also miscalculations about the rate at which nuclear power plants could be developed with minimal danger to health and safety. During the earlier days of nuclear power "planning," insufficient thought was given to the problem of thermal pollution—a problem which since has added to the difficulty of bringing nuclear power plants on stream with a minimum of delay.

The "energy crisis" was brought to a head in October 1973 by the latest outbreak of hostilities in the Middle East. Since then, it has been clear that an uninterrupted supply of Middle Eastern oil can no longer be taken for granted. Following the imposition of the oil embargo against the United States, there were various public pronouncements about the desirability of fuller utilization of domestic sources of energy—notably coal and shale oil.

Coal reserves in the United States are adequate to satisfy a large part of the nation's energy needs for many years. In West Virginia alone, for example, there are more than 90 billion tons of deep coal reserves. At an annual rate of consumption of 800 million tons—about 54 percent more than domestic consumption in 1972—this reserve alone would satisfy the nation's coal requirements for a century.[4] A major problem is that much of this huge stock of coal is relatively high in sulfur content. When burned in conventional boilers to generate electricity—or when used for industrial purposes—sulfur emissions would exceed those permitted by the Clean Air Act of 1970. There are several promising alternatives, however, and one is to convert coal to gas.

Coal Gasification

Why bother to convert coal to gas? Why not simply use gas in its natural state? Until recently, petroleum drillers considered natural gas to be "a waste product, something to be burned off in the field so they could get at the oil they sought."[5] This is no longer true. Lapp has estimated that by 1990 the projected demand for gas in the United States will amount to about 45 trillion cubic feet. The supply from 1970 proven reserves, and from potential new natural gas reserves, will amount to approximately 18 trillion cubic feet. Liquid gas imports plus Alaskan and synthetic gas could add another 9 trillion cubic feet. But this will leave a gap between requirements and supply of approximately 18 trillion cubic feet.[6] Even if these projections should be off by a very large percentage, they show that natural gas will account for a declining share of the nation's energy requirements. Lapp's projections highlight the urgent necessity for the development of economically feasible methods of converting coal to gas.

The idea of converting coal to gas is not a particularly new one. As early as 1691, Dr. John Clayton discovered that inflammable gas could be obtained by the distillation of coal. And between 1792 and 1802, William Murdoch "demonstrated the possibility of making gas from coal and using it as a lighting agent on a large scale."[7]

When coal is heated in the absence of air, gaseous hydrocarbons are produced. The remaining carbon is reacted by partial combustion to obtain carbon monoxide (CO) as coal is converted to a gaseous form. Sulfur is present in this gas as H_2S, but it is possible to remove this sulfur completely. There are two general goals for converting coal to gas. One is to develop a process that would yield gas with a high energy content per cubic foot (high-BTU gas), and the other a gas with relatively low energy content per cubic foot (low-BTU gas).

High-BTU Gas

High-BTU gas may be produced by using almost pure oxygen for the partial combustion of coal to convert carbon to carbon monoxide (CO) followed by conversion of the (CO) to methane (CH_4). One pilot plant for the production of high-BTU gas is located in Chicago on the same site used years ago by the Chicago Gas Company to manufacture illuminating gas. This pilot plant uses the "Hygas" process in which finely ground coal is mixed with a light oil, and the mixture is forced into a reactor. The oil and volatile matter rise to the top of the 140-foot reactor, while the coal particles fall down through an upward stream of carbon monoxide and steam. The steam and oxygen are pumped into the bottom of the reactor, and the interaction of steam and carbon monoxide with coal char produces methane, the principal ingredient of natural gas.[8]

Another pilot plant using a different process is being constructed near Pittsburgh. Plans have been made to locate plants in several western states. The objective of those engaged in research on the production of high-BTU gas is to develop processes to produce a synthetic substitute for natural gas with its high energy content (about 1,000 BTUs per cubic foot). A synthetic gas with high energy content can be transmitted long distances in pipelines just as natural gas is shipped today. It may be some time, however, before the production of high-BTU synthetic gas is technologically and economically feasible. Alternative processes now being developed could be used to produce low-BTU gas. This gas would not be transported long distances but would be consumed near or at the production site. There it could be used to generate electric energy. The electric energy could be transported over relatively long distances with existing technology.

Low-BTU Gas

The objective of research on low-BTU gas is to obtain at the lowest possible cost a *clean* gas from coal. Gas producers that made low- to medium-BTU, nitrogen-diluted gas were fairly common a half century ago. The gas thus manufactured had a number of objectionable qualities, however. It was a dirty gas, producing considerable amounts of dust, soot, and tar. The sulfur content of this gas was also objectionably high even fifty years ago. The old gas producers were small, and they had relatively high unit costs of production. As natural gas became available in increasing abundance, low-BTU gas production virtually disappeared in the United States.[9] Now that the available supply of natural gas is seen to be limited, there is considerable pressure once again to develop low-cost methods of producing clean, low-BTU gas from coal.

In the combustion process used to produce low-BTU gas, air is used instead of pure oxygen, and the resulting gas has a relatively high nitrogen content. This adds to the volume of the gas, which in turn makes it expensive to transport long distances. It could be used effectively, however, to generate electric energy at the site with relatively little loss of efficiency. The efficiency of converting the energy obtained from the combustion of fossil fuel to electricity could be increased substantially over the present 39 percent by the successful development of low-BTU gasification facilities in conjunction with advanced gas-turbine power generation systems.[10] It will be easier to remove sulfur from low-BTU gas than from stack gases after combustion.

There are a number of technically feasible processes for converting coal to low-BTU gas. None of these was economically feasible at the time this report was written. But as delivered prices of electric energy rise—as they almost certainly will—processes not considered feasible today may be used in the near future.[11]

Simulating the Production of Low-BTU Clean Gas in West Virginia

The actual construction of low-BTU, coal-conversion plants in West Virginia may be a number of years in the future. A considerable amount of work has been done, however, to estimate the cost of several conversion processes assuming that no new technological difficulties are encountered.[12] These studies are the primary sources of the *direct* cost estimates used in simulating the coal to gas conversion processes in West Virginia.

Two processes were selected for analysis as representative of the larger number now being considered. The selected processes are the *Dual-Stage, Fluidized-Bed Process* (West Virginia University) and the *Atmospheric Fixed-Bed Process* (Bureau of Mines).[13]

The fluidized-bed process generates only gas that would subsequently be used to produce electricity. The fixed-bed process would generate a by-product, ammonium sulphate (NH_4SO_4), for which a ready market is assumed.

Figure 6-1 shows a generalized scheme for producing low-BTU gas (130 to 170 BTUs per cubic foot) to be subsequently used for producing electrical power. To convert coal to gas, carbon must be converted to carbon monoxide, or hydrogen must be added to form a light gaseous hydrocarbon. This step takes place in the gasifier of Figure 6-1. Air may be compressed before entering the gasifier. Precursors of pollutants–such as sulfur–must be converted to removable compounds, and undesirable inorganic matter, such as ash, must be separated. Different operations, illustrated in Figure 6-1, are required to accomplish these steps. The final, clean low-BTU gas is burned, and electricity could be generated by expansion through large gas turbines (assuming these become technically feasible) or in conventional boiler operations.

Figure 6-2 shows the schematic flowsheet for producing gas using a stirred, fixed-bed system (Bureau of Mines). It can be operated at atmospheric pressure, or higher pressures, depending upon differences in size of equipment and the final electrical-generating equipment used. In a fixed-bed unit the fuel bed is supported by a grate. The coal moves down the bed toward the grate leaving a

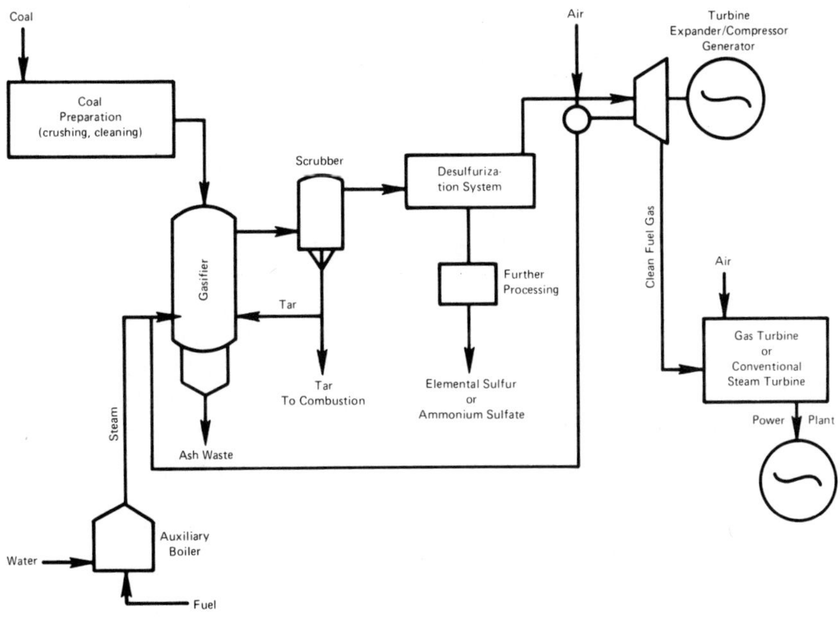

Figure 6-1. Simplified Proposed Gasification System–Power Plant Complex

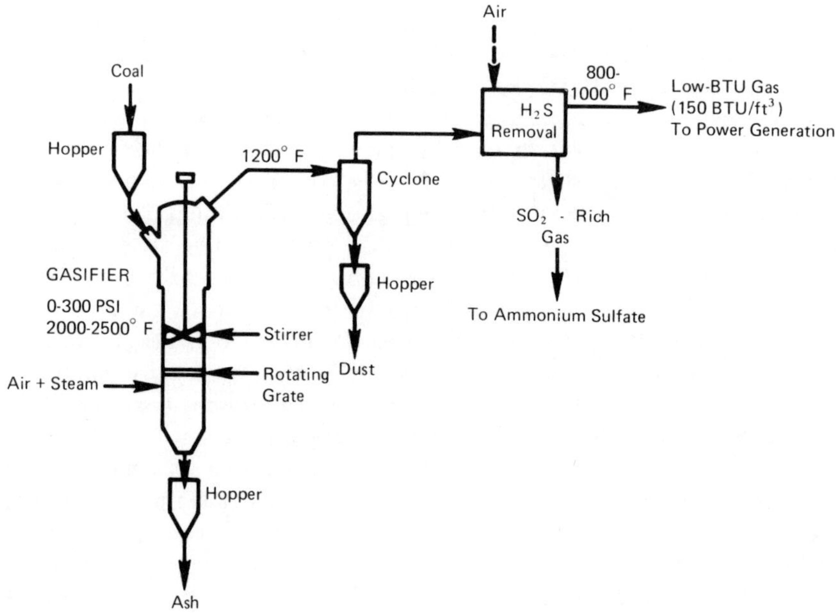

Figure 6-2. Fixed-Bed Gasification

residual ash to be discharged from the bottom. To use this process, the fuel bed must be stirred extensively to eliminate the formation of clinkers (agglomerates of incombustible matter) that are commonly associated with eastern United States coals.

In the proposed fluidized-bed system, agitation is maintained by a high gas flow that causes the solid and gas to act as a fluid. This allows for better thermal efficiency from increased contact between gas and recirculating solids.[14] It also increases the likelihood that the resulting gas could be burned directly in gas turbines. This process is illustrated by the modified flowsheet of Figure 6-3. The fluidized-bed gasifier we have simulated is a low-velocity system. The velocity of air (or gas) would be 0.66 feet per second in the simulated system, compared with a velocity of 10 or more feet per second in the system proposed by Squires.[15]

The simulated processes have been compared with a conventional coal-fired, steam-generating plant, which we assume has been fitted with an SO_2 flue-gas scrubbing system using limestone in the sulfur abatement process. The method used to estimate direct operating costs generally followed the procedures used by Katell et al. and by Wen cited above. Modifications were made when necessary to insure consistency among the three power-generating complexes. To

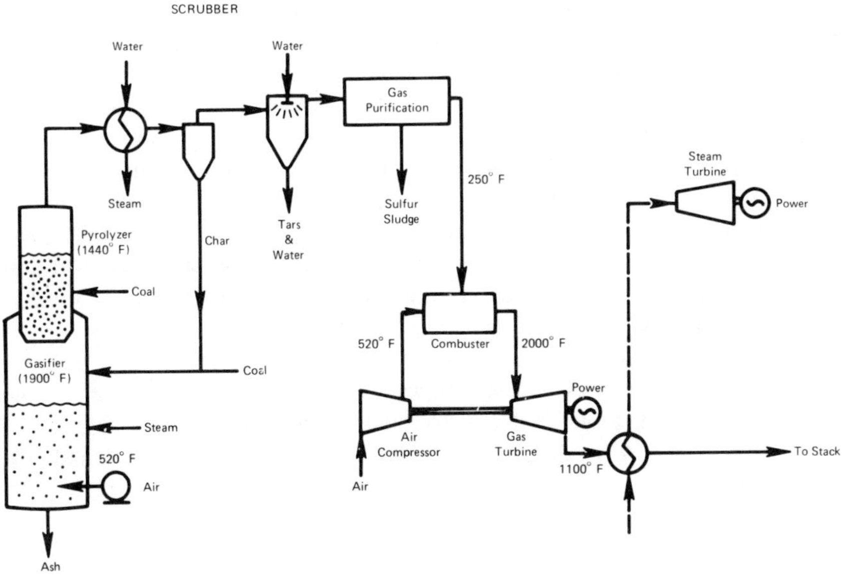

Figure 6-3. Dual-Stage Fluidized-Bed Gasification System

further insure consistency, we assumed that all complexes would use the same grade of Pittsburgh–seam West Virginia coal. Three coal-conversion power-generating complexes were simulated, each with a rated capacity of 2,500 megawatts. The estimation of capital costs in the simulated sectors also followed the methods used by Wen and by Katell, et al. For the conventional plant with an SO_2 scrubber system, capital and operating costs were estimated following the procedures used by Burchard and his associates.[16]

The following detailed assumptions were made about the operation of each of the complexes. We assumed that each plant would operate on the average of 85 percent of generating capacity for 90.4 percent of each year.[17] We assumed that the coal used would have a sulfur content of 2.8 percent, and that 85 percent of the sulfur would be removed in the conventional plant with SO_2 scrubber. The investment required to install a kilowatt of capacity was assumed to be as follows: for gas-turbine power equipment, \$70/kw; for conventional steam boiler power equipment, \$100/kw; and for the SO_2 flue-gas scrubber, \$48/kw.[18] Taxes and insurance were estimated at 6 percent of total investment, and the interest rate used was 7 percent. We assumed that ammonium sulfate would be sold at \$20 per ton. The cost of coal was estimated as the average 1972 price of coal, at the mine, in northern West Virginia, which was \$7.53 per ton.

Each of the simulated sectors was assumed to consist of three 2,500 megawatt complexes. The additional conventional plant was also assumed to be of the same size. The assumption of 7,500 megawatts of additional capacity is conservative, since an expansion of this magnitude would not quite double West Virginia's 1970 generating capacity. It is well within the limits of growth that might be expected on the basis of long-range projections of national energy needs made by the Federal Power Commission and by other public and private agencies. To use the West Virginia input-output model for simulation purposes, it was necessary to assume that these complexes would be completed and in operation by 1975. There is not even a remote possibility that technological developments would permit a single complex to be completed by that year, but this detracts from the realism of the simulation in this chapter only in terms of prices. As in the earlier part of this study, all value figures in the simulations are expressed in projected 1975 prices.

In making the direct cost estimates, we assumed that the capital costs for the gasification processes would be distributed in the same way as the capital coefficients in the West Virginia chemical sector.[19] The power plant and coal preparation sections of the complex were assumed to have the capital-cost distribution of the West Virginia electric utility sector.[20] Operating costs were distributed in the same way to obtain the technical (or input) coefficients. We assumed that all interindustry transactions, except in banking and insurance, occurred in West Virginia. Replacement capital expenditures were derived as in the original West Virginia simulation study.[21] Labor requirements, however, were taken from the estimates given by Wen in his June 1973 paper, and from Katell's 1973 ACS paper. The fluidized-bed gasification process is more complex than the others and requires more operating personnel. In the simulated gasification complexes, operating personnel requirements for the power plant portions of each complex were estimated from current requirements in conventional electric utility systems.

Comparative Analysis of the Simulated Coal-Gas Conversion Processes

The method used to simulate the coal gasification-power generation processes discussed in this section is identical to that used in the earlier West Virginia simulation study.[22] The simulation uses both the comparative static and dynamic models and is essentially an impact study. Impact studies are typically conducted within a static framework, and the output and employment effects derived below were obtained by using the West Virginia comparative static model. However, the dynamic model was also used—as in our earlier study—to estimate the time-phasing of impacts on all sectors during construction of the simulated complexes.

The input coefficients for the simulated sectors are given in Table 6-1. For the purposes of this study, we have combined the underground and surface coal

Table 6-1
Input Coefficients for Simulated Sectors

	(1) Dual-Stage Fluidized-Bed	(2) Atmospheric Fixed-Bed	(3) New Conv. Plant with SO_2 Scrubber
1. Agriculture			
2. Coal Mines (undgd.) 3. Coal (strip & auger)	.216662	.242857	.188286
4. Petroleum & natural gas			
5. All other mining			
6. Building contractors			
7. Non-building contractors			
8. Special contractors			
9. Food products (n.e.c.)			
10. Dairies			
11. Bakeries			
12. Beverages			
13. Apparel & accessories			
14. Logging & sawmills			
15. Furn. & wood fabrication			
16. Printing & publishing	.005154	.004286	.002820
17. Chemicals	.001501	.030508	.005900
18. Petroleum products			
19. Glass			
20. Stone & clay products	.004704	.003905	.006681
21. Primary metals	.000851	.000714	.000477
22. Fabricated metals	.011359	.009381	.006291
23. Machinery (except electric) 24. Electric machinery	.000650	.000524	.000347
25. Transportation equipment			
26. Instruments & products			
27. All other manufacturing			
28. Restaurants & bars	.001051	.000857	.000607
29. Wholesale trade	.008807	.007238	.004816
30. Retail food stores			
31. Auto service stations			
32. All other retail	.001051	.000857	.000607
33. Banking	.015412	.016190	.016529
34. Other finance			
35. Insurance agents	.002102	.002229	.002343
36. Real estate	.001234	.000952	.006551
37. All other FIRE			
38. Hotels & lodgings			
39. Medical & legal services	.001301	.001048	.000694
40. Educational services			
41. All other services	.007956	.006571	.004382
42. Railroads	.026020	.029286	.022560
43. Trucking & warehousing	.011926	.013952	.018438
44. All other transportation	.003970	.004381	.003384
45. Communications	.004804	.004873	.003905
46. Electrical systems			
47. Gas systems			
48. Water & sanitary services	.000083	.000095	.000087
49. Simulated sector	.009507	.037619	.038612
50. Households	.139755	.099333	.077657
51. Out-of-state (imports)	.157618	.165762	.148764
52. All other final payments	.366525	.316587	.439262
Total	1.000000	1.000000	1.000000

sectors. The expanded conventional plant would require less coal directly per dollar of output than either of the gasification processes. But as we will see below, when all direct *and indirect* effects are taken into account, the dual-stage fluidized-bed process would use the smallest amount of coal per dollar of output.

Another difference among the three simulated sectors, alluded to earlier, is the larger labor requirement of the fluidized-bed process. This is a more labor-intensive coal-gas conversion method than the other. The high labor-intensity of the fluidized-bed system is due in part to the fact that it is a low-velocity process in an early stage of development. It is also partly the result of our assumption that there would be no economies of scale on a sectoral basis.[23] In terms of labor efficiency, the conventional plant is more economical than either of the coal to gas conversion processes. But in an economic development context, labor-saving production processes are not necessarily considered to be advantageous. If a given process has relatively high labor requirements and a low *unit* cost of production, it would be given a high priority by development planners.

The simulated direct input coefficients were added to the West Virginia input-output model. The data on which these coefficients are based were originally given in 1972 or 1973 prices. These prices were deflated to 1965 dollars to make them compatible with the original (1965) West Virginia coefficients. A constant output of approximately 50.5 million megawatt hours was assumed for each simulated sector. But the cost of producing this amount of energy varied from sector to sector; hence, the *value* of final demand was also different for each sector.[24] The final demand values were inflated to 1975 prices using extrapolated deflators employed by the Council of Economic Advisors.[25] All subsequent value calculations were then expressed in 1975 dollars.

The impacts of the three simulated sectors on West Virginia's total gross output, in millions of 1975 dollars, are summarized in Table 6-2. The differences in impacts are explained by variation in costs of production and in interindustry linkages. These impacts on selected sectors are given in Figure 6-4, and the complete sectoral details are presented in Appendix Table 6A-1.

Table 6-2
Summary of Direct and Indirect Impacts of Simulated Sectors on the West Virginia Economy in 1975 Dollars

	Changes in Total Gross Output	
	Millions of 1975 Dollars	Percent
1. Dual Stage Fluidized-Bed	\$ 1,382.8	3.06%
2. Atmospheric Fixed-Bed	1,483.4	3.28
3. New Conventional Plant with SO_2 Scrubber	\$ 1,601.2	3.54%

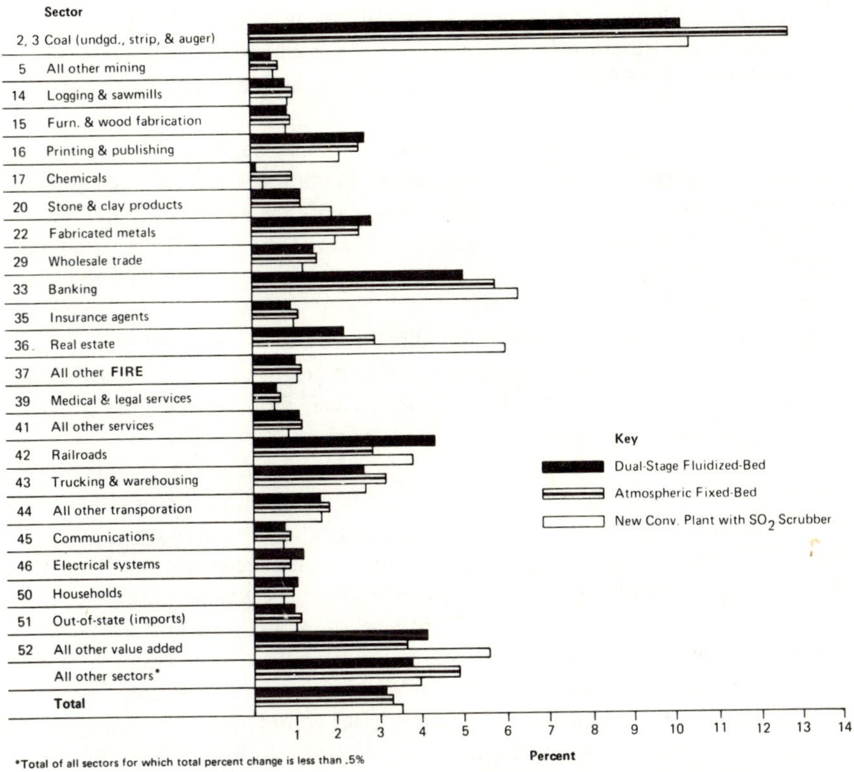

Figure 6-4. Employment Impacts on all Sectors of Simulated Sectors (Percent Change)

The fixed-bed process would have a larger overall impact on the state economy than the fluidized-bed. But additional conventional generating capacity, with SO_2 scrubbers, would have an even larger impact adding more than $1.6 billion to Gross State Product.

As expected, the sector that would benefit most is the coal industry. The atmospheric fixed-bed gas producer would require more coal to generate 50.5 million mwh of electric energy than the fluidized-bed system. Additional conventional plant capacity (with SO_2 scrubbers) would also require only about 82 percent as much coal as the atmospheric fixed-bed producer. Clearly, extracting energy from coal via the fixed-bed processes of gasification would be less "efficient" than the extension of conventional plant (with SO_2 scrubbers). The *quid pro quo*, however, is that there would be a major reduction in atmospheric pollution. On both efficiency and environmental grounds, however, the dual-stage fluidized-bed process would be the most desirable of the three simulated methods of adding to West Virginia's electricity generation capacity.

Although the fluidized-bed complex would buy less from coal and chemical sectors than the other gasification process and would have a smaller total impact on the state economy, it would provide more employment than the other simulated sectors.

The employment impacts of the simulated sectors are given in Table 6-3. The fluidized-bed process is the most labor-intensive, although this may be partly the result of our assumption of fixed input-output proportions. At the capacity level assumed, this process would generate 7,630 direct new jobs in West Virginia. And despite its low employment multiplier (1.89), it would generate the largest number of direct and indirect jobs, 14,394. The atmospheric fixed-bed producer ranks second. It would generate 5,866 direct jobs and 14,188 direct and indirect jobs. The smallest employment impact would be registered by the expansion of conventional capacity with SO_2 scrubbers. Here, 5,037 direct jobs would be created, and total new employment resulting from the expansion of conventional capacity would amount to only 12,127 jobs. From a manpower point of view, the fluidized-bed process would do the most for West Virginia. But either of the simulated gasification processes would have a significant employment impact on the regional economy. The employment impacts on selected sectors are given in Figure 6-5. The details for all sectors are presented in Appendix Table 6A-2.

The investment requirements of the three systems analyzed in this chapter are given in Table 6-4. The conventional method would require more than \$1.6 billion of additional capital (1975 prices) to expand capacity by 7,500 megawatts. The fixed bed system would require considerably less or about \$1.4 billion. The fluidized-bed system, which would cost less than \$1.3 billion, would be the most economical of the simulated systems in its capital requirements. This is also indicated by Column 3 of Table 6-4 where capital output ratios for the three methods are summarized. The conventional system has the highest ratio and the lowest is that for the fluidized-bed system. When capital requirements are expressed per worker, as in Columns 4 and 5 of this table, the rankings are not altered.

Finally, we come to the definitive criterion for selecting a production process when several alternatives are available, *the unit cost of production.* In the case of utilities, this cost, plus the mark-up allowed by state public utility commissions as an acceptance rate of return, represents the average price to consumers.[26]

Column 1 of Table 6-5 gives the estimated value of sales, excluding the return on original investment, in 1975 dollars. The second column gives the (constant) volume of sales in kwh. Column 3 gives the cost of production expressed in cents per kwh.[27] The fluidized-bed system would have a substantially lower unit cost of production than either the fixed-bed or expanded conventional systems. It might be added that since the fluidized-bed system also would have the lowest investment requirements of the three systems compared, the difference in average consumer prices between this system and the other two would be even larger.

Table 6-3
Employment Impact of Simulated Sectors

	Change in TGO (Millions of 1975 $)	Employment: Simulated Sector[a]	Employment: All Sectors[b]	Employment Multiplier[c]	TGO per Total Employment[d] (Thousands of 1975 $)
1. Dual-Stage Fluidized Bed	$1,382.8	7,630	14,394	1.89	$ 96.1
2. Atmospheric Fixed-Bed	1,483.4	5,866	14,188	2.42	104.6
3. New Conventional Plant with SO_2 Scrubber	1,601.2	5,037	12,127	2.41	132.0

[a]Direct and indirect employment in simulated sector.
[b]Direct and indirect employment in all sectors.
[c]All sectors employment divided by employment in the simulated sector.
[d]Change in TGO divided by all sectors employment.

Table 6-4
Investment Requirements of Simulated Sectors

	Investment (Thousands of 1975 $)	Output (Thousands of 1975 $)	Capital Output Ratio	Capital Requirements per Worker: Simulated Sector	Capital Requirements per Worker: All Sectors
1. Dual-Stage Fluidized-Bed	$1,264,899	$665,411	1.90	165,780	87,877
2. Atmospheric Fixed-Bed	1,407,900	719,692	1.96	240,010	99,232
3. New Conventional Plant with SO_2 Scrubber	$1,632,060	$790,762	2.06	324,014	134,581

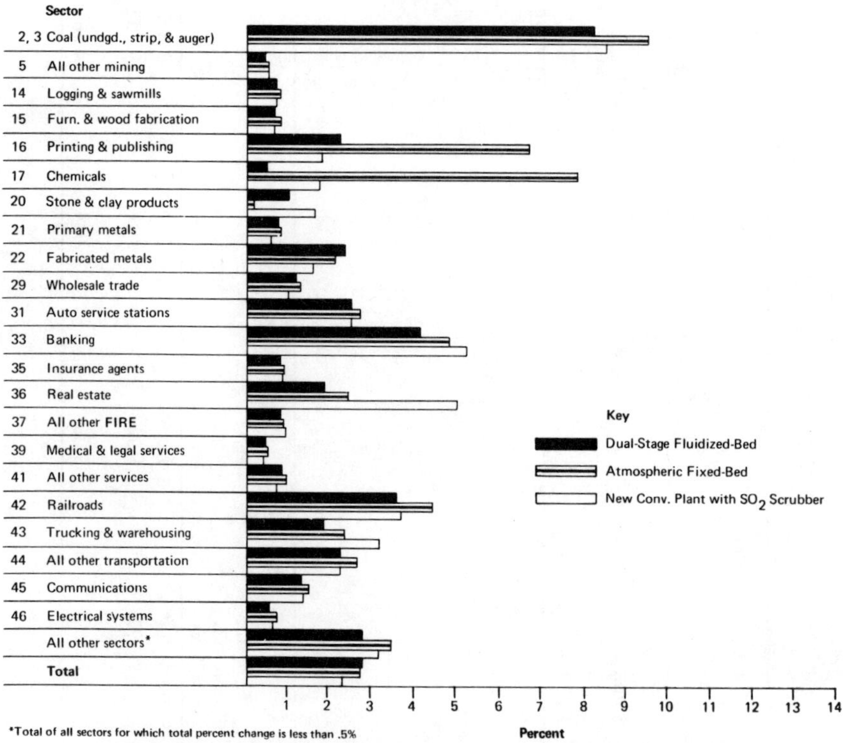

Figure 6-5. Employment Impacts on All Sectors of Simulated Sectors (Percent Change)

The Dynamic Analysis

Planning and other preparatory work on the construction of a major gasification-generation complex would start several years before the anticipated completion date. Most of the construction work would be done over a period of about three years, and the largest impacts on the state economy would be felt during this time interval. The time-phasing of investment impacts can be traced by using a dynamic input-output model in its "dynamic inverse" form.[28] The dynamic inverse illustrates the historical "unfolding" of a new investment. We have used the West Virginia dynamic inverse to show what the sectoral impact would be over a period of several years if a gasification-generation complex were to be completed in 1975.

In all three cases the major impacts before the target year would be on the construction sector; and the pattern is similar in all three cases. The peak impact on state output would be felt in the year prior to completion of the complex. This makes sense intuitively because it is the year in which all construction

Table 6-5
Unit Cost of Production of Simulated Systems

	Direct and Indirect 1975 $ Sales (Thousands of 1975 $)	Kilowatt Hours (Thousands)	Cost of Production (Cents per kwh)
1. Dual-Stage Fluidized-Bed	$665,411	50,483,880	1.38¢
2. Atmospheric Fixed-Bed	719,692	50,483,880	1.43
3. New Conventional Plant with SO_2 Scrubber	$790,762	50,483,880	1.57¢

Note: Figures exclude return on investment.

would be completed and equipment would be installed. During the first year of operation there would be a major upsurge in the sale of coal. Other things remaining unchanged, however, one would expect a substantial drop in construction activity. This is what the dynamic simulation of the complexes shows. It also shows that much of the demand for the output of other sectors during the construction phase would be a derived demand resulting from the expansion of construction activity. In the target year—again assuming all other things remain unchanged—most sectors would register declines in total gross output. The impact of each simulated system on selected sectors is shown graphically in Figures 6-6 through 6-8.

In the two simulated gasification-generation complexes, increases in output in the target year outweigh declines so that the total impact is positive although much smaller than that in the prior year. In the simulated expansion of conventional capacity with SO_2 scrubbers, however, negative changes in the target year outweigh the increases in output that are concentrated in coal and a few other sectors so that the total impact on gross state output would be negative in that year. It is important to stress that the dynamic model shows *changes* in annual output rather than the level of output in each year. Thus, negative figures in any given year indicate a decline in output. While this might be a fairly large number in absolute terms, it is meaningful only in relation to the level of output in the prior year. The details of the annual changes in output by sector for each of the simulated complexes are given in Appendix Tables 6A-3 through 6A-5.

Summary and Conclusions

The successful implementation of a set of coal gasification-power generation complexes would have a major impact on the West Virginia economy. On the basis of the assumptions set out earlier in this chapter—assumptions that square realistically with long-range projections of the nation's energy needs—this

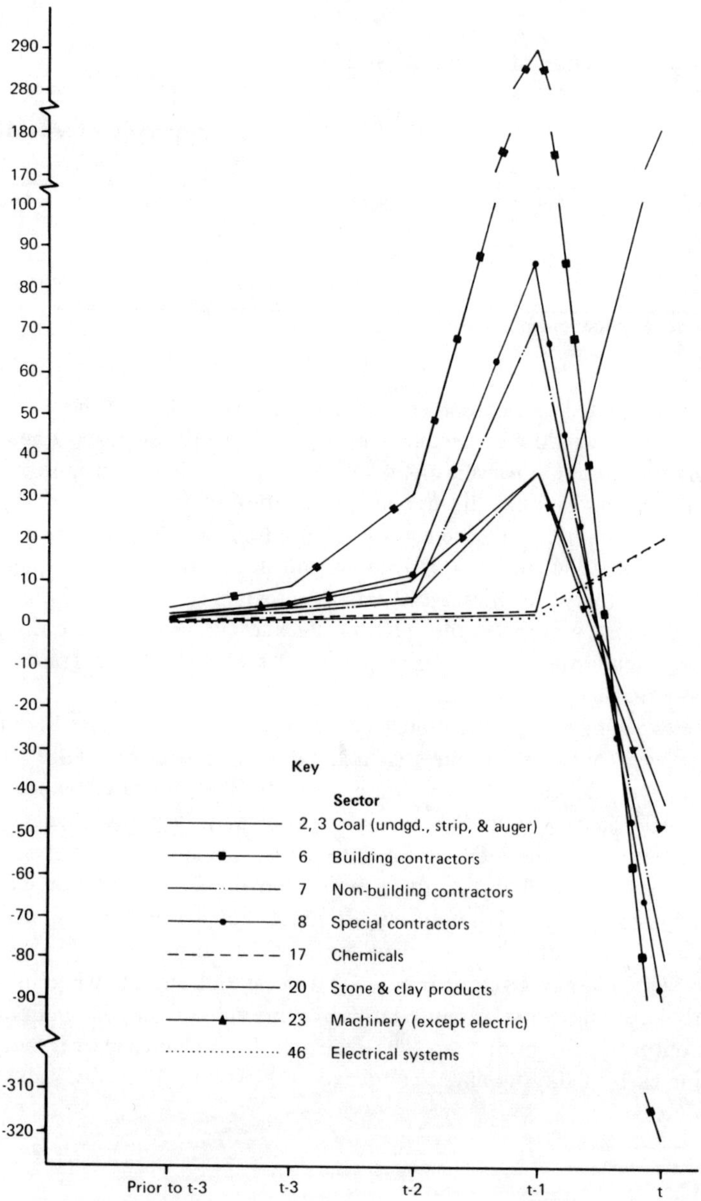

Figure 6-6. Direct and Indirect Impact on Selected Sectors of Dual-Stage Fluidized-Bed System in Target and Preceding Years (Thousands of 1975 Dollars)

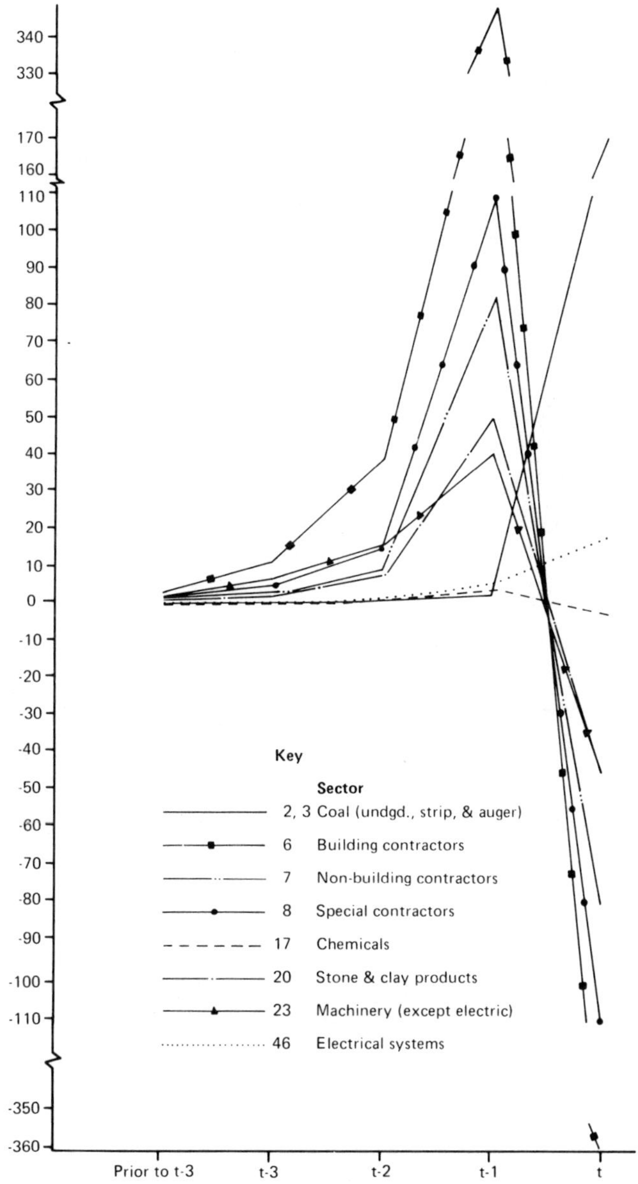

Figure 6-7. Direct and Indirect Impact on Selected Sectors of Atmospheric Fixed-Bed Process in Target and Preceding Years (Thousands of 1975 Dollars)

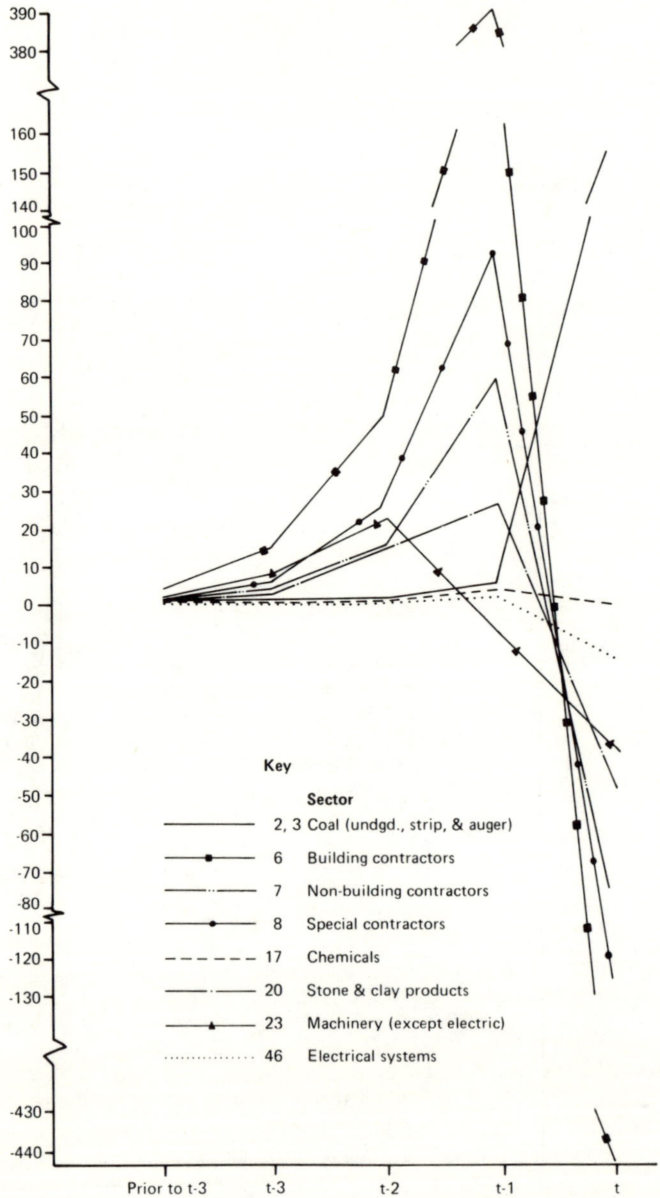

Figure 6-8. Direct and Indirect Impact on Selected Sectors of Conventional Plant with SO_2 Scrubber in Target and Preceding Years (Thousands of 1975 Dollars)

development would add about $1.4 billion to gross state product in 1975 prices. Direct employment would range from about 5,900 new jobs to more than 7,600 jobs. Total employment, however, would go up by more than 14,000 jobs when all indirect effects are taken into account. West Virginia's coal industry, which has been lagging in recent years, would be the principal industrial beneficiary of such a development. But the ramifications of substantial investment in low-BTU gasification would be felt throughout the state economy, and there would be significant over-spill benefits in states that sell capital equipment to West Virginia.

Of the two processes examined in this chapter, the fluidized-bed method is preferable on economic and conservation grounds. It would use less coal per unit of output than the fixed-bed method; it would require a smaller initial investment; it would be more labor-intensive and hence provide more jobs; and it would produce electricity at a lower unit cost than competing methods.

The electric energy sector in West Virginia will continue to grow. But if capacity is expanded without substantial abatement of air pollution, the favorable economic impact will be offset to some extent by further environmental deterioration. The development of effective SO_2 scrubbers would, of course, permit the expansion of conventional capacity with less additional environmental damage than that caused by the growth of this sector in the recent past. But the great advantage of either of the gasification processes analyzed in this chapter is that they would make a major contribution to the economic development of the state with no adverse environmental consequences. At first glance this might appear to violate the "no free lunch" principle stressed by environmental economists. But life is really never that simple. The *quid pro quo* in the case of coal gasification is the need for a major research effort to solve remaining technical problems. In the words of Arthur Squires, ". . . good technologies for providing clean power gas will be developed only if a great deal more talent is recruited for the work, and probably only with a large input of Federal money."[29]

Appendix 6A

Table 6A-1
Impact of Simulated Sectors on West Virginia Total Gross Output by Sector (Thousands of 1975 Dollars)

	(1) Dual-Stage Fluidized-Bed		(2) Atmospheric Fixed-Bed		(3) New Conv. Plant with SO_2 Scrubber	
	Change in TGO		Change in TGO		Change in TGO	
	Thousands of dollars	Per cent change	Thousands of dollars	Per cent change	Thousands of dollars	Per cent change
1. Agriculture	127	.06	143	.07	130	.06
2. Coal mines (undgd.) 3. Coal (strip & auger)	152,272	10.05	191,039	12.61	157,326	10.38
4. Petroleum & natural gas	252	.16	513	.33	311	.20
5. All other mining	513	.45	682	.60	647	.57
6. Building contractors	259	.05	371	.07	702	.13
7. Non-building contractors	502	.11	668	.14	682	.14
8. Special contractors	1,466	.33	1,920	.44	1,878	.43
9. Food products (n.e.c.)	229	.25	261	.28	219	.24
10. Dairies	58	.04	60	.04	49	.04
11. Bakeries	34	.04	34	.04	27	.03
12. Beverages	15	.03	15	.03	12	.02
13. Apparel & accessories	31	.06	35	.07	28	.06
14. Logging & sawmills	996	.78	1,155	.91	1,047	.82
15. Furn. & wood fabrication	384	.77	437	.88	401	.80
16. Printing & publishing	5,229	2.63	5,020	2.52	4,142	2.08
17. Chemicals	1,330	.05	24,242	.93	5,339	.20
18. Petroleum products	606	.38	722	.45	735	.46
19. Glass	25	*	40	.01	26	*
20. Stone & clay products	3,462	1.19	3,281	1.13	5,475	1.89
21. Primary metals	1,630	.09	1,730	.10	1,235	.07
22. Fabricated metals	7,346	2.79	6,707	2.54	5,122	1.94
23. Machinery (except electric)	244	.11	295	.13	251	.11
24. Electric machinery	478	.20	469	.20	332	.14

Table 6A-1 (cont.)

	(1) Dual-Stage Fluidized-Bed		(2) Atmospheric Fixed-Bed		(3) New Conv. Plant with SO_2 Scrubber	
	Change in TGO		Change in TGO		Change in TGO	
	Thousands of dollars	Per cent change	Thousands of dollars	Per cent change	Thousands of dollars	Per cent change
25. Transportation equipment	118	.04	144	.05	125	.04
26. Instruments & products	13	.07	52	.30	18	.10
27. All other manufacturing	447	.17	656	.26	468	.18
28. Restaurants & bars	1,067	.37	1,037	.36	840	.29
29. Wholesale trade	11,687	1.42	12,468	1.51	9,782	1.19
30. Retail food stores	11	*	15	*	9	*
31. Auto service stations	316	.29	357	.33	327	.30
32. All other retail	2,130	.21	2,256	.22	1,933	.19
33. Banking	13,892	4.92	16,095	5.70	17,566	6.22
34. Other finance	316	.22	438	.30	309	.21
35. Insurance agents	6,234	.86	7,442	1.02	7,084	.97
36. Real estate	3,424	2.15	4,555	2.86	9,379	5.89
37. All other FIRE	1,204	.93	1,423	1.10	1,326	1.03
38. Hotels & lodgings	92	.09	134	13	109	.10
39. Medical & legal services	2,538	.52	2,790	.57	2,256	.46
40. Educational services	33	*	42	*	35	*
41. All other services	11,187	.99	12,165	1.08	9,304	.83
42. Railroads	23,043	4.24	28,265	5.20	23,804	4.38
43. Trucking & warehousing	11,575	2.14	14,863	2.75	20,126	3.72
44. All other transportation	3,891	2.58	4,734	3.14	3,965	2.63
45. Communications	5,412	1.57	6,172	1.79	5,462	1.59
46. Electrical systems	4,005	.69	4,727	.82	4,213	.73
47. Gas systems	916	.08	1,480	.13	1,122	.10
48. Water & sanitary services	244	.27	320	.36	296	.33
49. Simulated sector	665,411		719,692		790,762	
50. Households	92,118	1.13	68,800	.85	59,038	.73
51. Out-of-state (imports)	102,447	.99	113,212	1.09	111,521	1.07
52. All other value added	241,591	4.05	219,273	3.69	333,941	5.59
Total	1,382,850	3.06	1,483,446	3.28	1,601,236	3.54

*Less than .01%

[a] Direct and indirect changes in TGO due to introduction of simulated activity.

Table 6A-2
Employment Impacts on All Sectors of Simulated Sectors

	(1) Dual-Stage Fluidized-Bed			(2) Atmospheric Fixed-Bed			(3) New Conv. Plant with SO_2 Scrubber		
	Changes in			Changes in			Changes in		
	Man hours	Employ-ment	Per cent change	Man hours	Employ-ment	Per cent change	Man hours	Employ-ment	Per cent change
1. Agriculture	862	*	0	973	*	0	882	*	0
2. Coal mines (undgd.)	6,707,141	3,001	8.33	7,589,738	3,517	9.76	6,930,328	3,101	8.61
3. Coal (strip & auger)	1,479,284	718	28.67	2,427,672	1,178	47.04	1,528,000	742	29.63
4. Petroleum & natural gas	18,564	9	.17	37,747	18	.34	22,912	11	.21
5. All other mining	12,443	6	.48	16,575	7	.56	15,715	7	.56
6. Building contractors	9,817	5	.05	14,056	7	.07	26,667	13	.13
7. Non-building contractors	17,189	9	.11	22,089	11	.14	23,326	12	.15
8. Special contractors	74,735	40	.34	97,919	52	.44	95,818	51	.43
9. Food products (n.e.c.)	8,531	4	.25	9,772	5	.31	8,224	4	.25
10. Dairies	2,072	1	.04	2,141	1	.04	1,734	1	.04
11. Bakeries	2,327	1	.04	2,286	1	.04	1,853	1	.04
12. Beverages	904	*	0	887	*	0	719	*	0
13. Apparel & accessories	7,511	4	.06	8,608	4	.06	7,046	4	.06
14. Logging & sawmills	127,044	61	.79	147,403	70	.91	133,474	64	.83
15. Furn. & wood fabrication	34,186	16	.75	38,990	19	.89	35,681	17	.79
16. Printing & publishing	351,329	177	2.63	337,290	170	7.94	278,302	140	2.08
17. Chemicals	28,425	12	.50	518,155	222	9.27	114,131	49	2.05
18. Petroleum products	6,742	3	.35	8,040	4	.48	8,171	4	.48
19. Glass	1,625	1	.01	2,645	1	.01	1,710	1	.01
20. Stone & clay products	172,152	86	1.19	163,164	82	.14	272,242	136	1.89
21. Primary metals	55,131	27	.88	58,515	29	.94	41,792	21	.68
22. Fabricated metals	566,979	256	2.79	517,653	234	2.55	395,315	178	1.94
23. Machinery (except electric)	14,020	8	.14	16,946	8	.14	14,452	7	.12
24. Electric machinery	23,582	11	.20	23,141	11	.20	16,408	8	.15
25. Transportation equipment	3,562	2	.04	4,328	2	.04	3,733	2	.04

Table 6A-2 (cont.)

	(1) Dual-Stage Fluidized-Bed			(2) Atmospheric Fixed-Bed			(3) New Conv. Plant with SO_2 Scrubber		
	Changes in			Changes in			Changes in		
	Man hours	Employment	Per cent change	Man hours	Employment	Per cent change	Man hours	Employment	Per cent change
26. Instruments & products	825	*	0	3,370	2	.03	1,175	1	.17
27. All other manufacturing	25,623	13	.18	37,588	19	.26	26,798	13	.18
28. Restaurants & bars	86,474	48	.37	84,075	46	.35	68,087	37	.29
29. Wholesale trade	632,852	301	1.42	675,129	321	1.51	529,693	252	1.19
30. Retail food stores	604	*	0	807	*	0	533	*	0
31. Auto service stations	23,148	13	2.98	26,145	14	3.21	23,983	13	2.98
32. All other retail	174,396	95	.21	184,741	101	.22	158,299	87	.19
33. Banking	524,028	289	4.92	607,126	335	5.71	662,640	365	6.22
34. Other finance	10,753	6	.23	14,869	8	.30	10,481	6	.23
35. Insurance agents	100,444	57	.86	119,899	68	1.02	113,330	64	.96
36. Real estate	122,121	70	2.15	162,464	93	2.86	334,475	192	5.90
37. All other FIRE	26,181	14	.93	30,962	17	.99	28,841	16	1.06
38. Hotels & lodgings	10,102	5	.10	14,770	7	.14	11,901	5	.10
39. Medical & legal services	334,730	183	.52	368,112	202	.57	297,606	163	.46
40. Educational services	4,591	2	0	5,738	3	.01	4,902	2	0
41. All other services	729,331	376	.99	793,089	409	1.08	606,532	313	.83
42. Railroads	981,962	450	4.24	1,204,568	553	5.21	1,014,399	465	4.38
43. Trucking & warehousing	407,624	177	2.14	523,396	227	2.75	708,708	307	3.72
44. All other transportation	76,508	43	2.60	93,069	52	3.14	77,958	44	2.66
45. Communications	230,964	114	1.57	263,404	130	1.79	233,091	115	1.59
46. Electrical systems	79,633	40	.69	94,008	48	.82	83,774	43	.74
47. Gas systems	7,634	4	.08	12,343	6	.12	9,356	5	.10
48. Water & sanitary services	12,697	6	.26	16,570	8	.34	15,342	8	.34
49. Simulated sector	15,030,445	7,630		11,555,148	5,866		9,922,668	5,037	
Total	29,359,827	14,394	3.23	28,958,123	14,188	3.18	24,923,207	12,127	2.72

Table 6A-3

Direct and Indirect Effects on Output in Target and Preceding Years of Simulated Dual-Stage Fluidized-Bed System (Thousands of 1975 Dollars)

	Prior to t-3	t-3	t-2	t-1	t
1. Agriculture	16	61	279	-105	-203
2. Coal mines (undgd.) 3. Coal (strip & auger)	35	134	557	1,600	186,414
4. Petroleum & natural gas	13	45	182	1,407	-1,156
5. All other mining	279	1,064	5,053	-10,046	969
6. Building contractors	2,881	8,007	30,381	290,178	-321,223
7. Non-building contractors	423	1,710	5,389	69,950	-76,313
8. Special contractors	940	3,635	11,211	84,337	-97,605
9. Food products (n.e.c.)	3	13	31	216	15
10. Dairies	1	3	15	72	41
11. Bakeries	0	3	10	42	29
12. Beverages	0	1	4	19	13
13. Apparel & accessories	0	3	9	51	34
14. Logging & sawmills	231	857	4,539	-6,401	717
15. Furn. & wood fabrication	212	721	5,125	-9,432	2,133
16. Printing & publishing	55	228	874	2,631	892
17. Chemicals	35	140	556	3,272	19,837
18. Petroleum products	89	364	1,229	13,210	-14,377
19. Glass	7	32	123	121	-266
20. Stone & clay products	418	1,624	5,824	36,363	-40,242
21. Primary metals	588	2,145	10,734	6,579	-18,978
22. Fabricated metals	349	1,330	4,456	17,758	-17,634
23. Machinery (except electric)	850	3,507	10,243	37,459	-49,240
24. Electric machinery	124	426	2,077	3,272	-4,827
25. Transportation equipment	15	73	0	663	-550
26. Instruments & products	4	16	61	1,009	-1,002
27. All other manufacturing	185	842	1,009	8,548	-9,716
28. Restaurants & bars	9	38	152	1,256	-796
29. Wholesale trade	272	1,023	4,044	8,940	-4,746
30. Retail food stores	0	0	3	12	-6
31. Auto service stations	15	57	212	931	-1,006
32. All other retail	407	1,588	5,776	-11,835	5,228
33. Banking	44	169	602	3,561	7,376
34. Other finance	6	23	92	469	-291
35. Insurance agents	80	311	1,142	6,259	-2,505
36. Real estate	36	142	530	2,873	-339
37. All other FIRE	13	53	204	806	-77
38. Hotels & lodgings	6	20	77	288	-308
39. Medical & legal services	28	108	420	2,253	-895
40. Educational services	1	4	19	26	-19
41. All other services	96	381	1,442	6,891	-145
42. Railroads	23	85	302	996	19,192
43. Trucking & warehousing	79	311	1,139	5,252	3,850
44. All other transportation	25	98	387	970	2,225
45. Communications	39	153	593	2,587	-1,854
46. Electrical systems	67	269	1,073	3,946	454
47. Gas systems	44	178	686	3,876	-3,736
48. Water & sanitary services	86	324	1,234	11,235	-12,666
49. Simulated sector					665,621
50. Households					
51. Out-of-state (imports)					
52. All other final payments					
Total	9,129	31,991	120,100	604,365	232,319

[a] Assuming final sales of $659,475 in year t.

Table 6A-4
Direct and Indirect Effects on Output in Target and Preceding Years of Simulated Coal to Gas Conversion Atmospheric Fixed-Bed Process (Thousands of 1975 Dollars)

	Prior to t-3	t-3	t-2	t-1	t
1. Agriculture	24	90	446	-59	-236
2. Coal mines (undgd.) 3. Coal (strip & auger)	53	197	823	2,201	173,143
4. Petroleum & natural gas	18	74	281	1,734	-1,475
5. All other mining	408	1,535	10,413	-10,811	1,061
6. Building contractors	3,179	11,808	39,002	348,648	-361,968
7. Non-building contractors	624	2,439	8,819	82,694	-79,301
8. Special contractors	1,388	5,318	15,444	114,086	-110,423
9. Food products (n.e.c.)	5	21	47	279	-42
10. Dairies	2	5	20	90	-45
11. Bakeries	2	3	14	54	-30
12. Beverages	0	2	5	24	-15
13. Apparel & accessories	0	3	12	66	-41
14. Logging & sawmills	336	1,247	7,092	-7,221	645
15. Furn. & wood fabrication	309	996	8,369	-10,554	2,313
16. Printing & publishing	81	341	1,260	3,366	1,530
17. Chemicals	53	203	884	3,986	-2,811
18. Petroleum products	132	522	1,946	15,939	-15,206
19. Glass	12	47	176	204	-326
20. Stone & clay products	612	2,349	8,531	49,481	-44,378
21. Primary metals	878	3,080	15,218	14,216	-19,583
22. Fabricated metals	513	1,962	6,954	27,743	-18,243
23. Machinery (except electric)	1,236	5,567	16,356	46,386	-45,926
24. Electric machinery	189	641	2,501	2,621	-4,424
25. Transportation equipment	23	113	-11	620	-530
26. Instruments & products	6	24	78	1,101	-1,112
27. All other manufacturing	272	1,287	1,769	11,637	-11,105
28. Restaurants & bars	14	57	204	1,542	-836
29. Wholesale trade	399	1,511	6,324	12,848	-6,188
30. Retail food stores	0	2	3	15	-9
31. Auto service stations	21	84	311	1,262	-1,151
32. All other retail	594	2,382	8,177	-13,221	5,090
33. Banking	65	249	854	4,532	6,975
34. Other finance	9	35	138	582	-399
35. Insurance agents	117	461	1,682	7,854	-3,426
36. Real estate	54	209	743	3,647	-1,194
37. All other FIRE	21	78	303	1,059	-212
38. Hotels & lodgings	8	32	114	381	371
39. Medical & legal services	41	159	608	2,850	-1,095
40. Educational services	2	6	26	30	-23
41. All other services	143	557	2,159	8,687	-251
42. Railroads	33	126	450	1,293	18,117
43. Trucking & warehousing	116	459	1,757	6,791	2,390
44. All other transportation	38	146	567	1,244	2,061
45. Communications	57	225	858	3,290	-2,295
46. Electrical systems	99	398	1,599	5,168	-156
47. Gas systems	65	261	1,031	4,967	-4,425
48. Water & sanitary services	129	479	1,598	13,532	-14,282
49. Simulated sector					719,732
50. Households					
51. Out-of-state (imports)					
52. All other final payments					
Total	12,380	47,790	175,955	766,884	180,266

a/ Assuming final sales of $692,677 in year t.

Table 6A-5
Direct and Indirect Effects on Output in Target and Preceding Years of Simulated Conventional Plant with SO_2 Scrubber (Thousands of Dollars)

	Prior to t-3	t-3	t-2	t-1	t
1. Agriculture	32	123	443	126	-646
2. Coal mines (undgd.) 3. Coal (strip & auger)	71	266	1,090	5,264	153,624
4. Petroleum & natural gas	24	96	448	1,422	-1,654
5. All other mining	546	1,978	11,107	-13,691	1,142
6. Building contractors	4,216	15,953	47,280	390,552	-445,835
7. Non-building contractors	842	3,125	16,886	57,536	-75,630
8. Special contractors	1,846	6,898	27,123	92,629	-124,957
9. Food products (n.e.c.)	6	26	67	264	-146
10. Dairies	1	6	22	94	-88
11. Bakeries	1	5	14	58	-56
12. Beverages	0	1	6	27	-26
13. Apparel & accessories	1	5	18	61	-62
14. Logging & sawmills	446	1,792	6,146	-7,966	346
15. Furn. & wood fabrication	408	1,536	7,666	-11,798	2,445
16. Printing & publishing	107	448	1,360	3,138	-930
17. Chemicals	70	266	1,267	3,997	-112
18. Petroleum products	178	680	3,445	11,551	-15,050
19. Glass	16	61	184	197	-418
20. Stone & clay products	819	3,222	14,331	34,307	-45,322
21. Primary metals	1,168	4,333	19,950	178,296	-201,440
22. Fabricated metals	678	2,611	12,373	12,131	-22,024
23. Machinery (except electric)	1,656	7,253	24,851	-1,814	-27,080
24. Electric machinery	248	814	1,930	14,374	-15,989
25. Transportation equipment	30	109	37	429	-430
26. Instruments & products	8	27	82	1,507	-1,539
27. All other manufacturing	374	1,453	2,987	9,184	-12,733
28. Restaurants & bars	18	75	259	1,626	-1,478
29. Wholesale trade	533	2,008	7,208	8,800	-9,534
30. Retail food stores	0	1	5	14	-13
31. Auto service stations	29	112	421	997	-1,336
32. All other retail	805	3,094	5,611	-11,555	6,094
33. Banking	86	330	1,146	4,334	7,430
34. Other finance	11	46	174	528	-530
35. Insurance agents	157	610	2,197	7,317	-4,962
36. Real estate	72	277	918	3,555	2,190
37. All other FIRE	27	104	389	1,030	-555
38. Hotels & lodgings	10	42	142	341	-453
39. Medical & legal services	54	213	778	2,954	-2,384
40. Educational services	1	8	26	144	-149
41. All other services	190	738	2,856	8,475	-5,112
42. Railroads	45	166	598	2,462	-14,702
43. Trucking & warehousing	154	608	2,514	5,862	6,005
44. All other transportation	50	194	718	2,606	-251
45. Communications	77	301	1,062	3,496	-3,379
46. Electrical systems	133	530	2,024	5,872	-3,090
47. Gas systems	86	347	1,501	4,066	-5,088
48. Water & sanitary services	170	646	1,934	15,208	-17,670
49. Simulated sector					790,885
50. Households					
51. Out-of-state (imports)					
52. All other final payments					
Total	16,500	63,537	233,594	850,007	-88,604

[a] Assuming final sales of $760,348 in year t.

7 Environmental Management and Regional Economic Development

This is not a lengthy report. Because of the analytical method used, however, it necessarily includes many details. Thus is might be helpful to summarize the objective, method, and general findings of the study.

Public policies may, at times, have contradictory objectives. In the late 1930s, for example, the nation adopted an agricultural policy of reducing excess production by limiting acreage. At the same time, another phase of agricultural policy had as its objective the development of new methods to increase crop yields. As these improved methods were adopted, there were substantial increases in agricultural productivity. The policy of reducing acreage was offset to some extent by the policy of increasing productivity. One objective of the present study was to determine whether or not the nation's regional development programs and its environmental protection programs might be at least partly in conflict with one another.

Laws designed to protect the environment could inhibit regional economic development if industrialists in a given region chose to relocate rather than invest in abatement equipment at their present locations. Should this happen in West Virginia's primary metals and chemical industries, for example, even relatively modest annual declines in final sales would have a severe impact on the state economy. A reduction of 1 percent a year in the final sales of the primary metals and chemical sectors would lead to a cumulative contraction over a period of 10 years of 38 percent in chemicals and 32 percent in primary metals. And there would be a cumulative reduction in investment in West Virginia of almost 30 percent, other things remaining unchanged.[1]

Fortunately, we found no evidence of managerial behavior that would inhibit regional economic development, although one power plant in the state has been closed and another is scheduled to be closed later. The continued opening of new and much larger plants, however, will add to this sector's total capacity. A few small foundries and other industrial establishments no doubt will be unable to comply with clean air standards and will close. But in the state's major, heavy-polluting sectors there is no evidence that sales or capacity will be reduced because of clean air regulations. There appears to be no conflict between the objectives of regional development policy and clean air legislation.

This study was devoted primarily to measurement of the direct and indirect costs of compliance with clean air standards. It involved collecting data for the calculation of new, direct input coefficients for 16 of the 48 basic sectors in the West Virginia input-output model.[2]

The adjustment of direct input coefficients to take into account the cost of installing pollution abatement equipment was a two-stage process. First, we estimated the costs of particulate abatement in the 16 sectors that will be affected by particulate regulations. The second step was to estimate the costs of installing equipment to reduce sulfur dioxide emissions to the limits permitted by clean air regulations. The individual estimates were later combined to provide estimates of the total operating costs of meeting clean air standards.

The increases in operating costs resulting from clean air legislation are relatively small, but they will have to be met on an annual basis. In absolute terms, however, changes in annual operating costs are dwarfed by the investment requirements of meeting clean air standards.

Both the expansion and replacement capital coefficients of the West Virginia dynamic model were adjusted to reflect the changes in investment that will result from compliance with clean air legislation. After all of the adjustments had been made in the input-output coefficients, we were in a position to estimate the costs of compliance with the primary and secondary standards established by the Clean Air Act of 1970.

The cost of clean air in West Virginia was estimated by calculating the direct and indirect production—before and after the installation of pollution abatement equipment—that would be required by each sector to produce a constant volume of goods and services for sales to final demand. We were attempting to estimate the value of the resources that would be required to produce a constant level of *final* output in each of West Virginia's 48 sectors, if the requirements of the Clean Air Act are to be met. We have called this estimate the "real" cost of pollution abatement.[3]

The costs of meeting particulate and sulfur dioxide standards were calculated separately and are reported in Chapters 3 and 4, but only the total costs of meeting clean air standards will be summarized here. There is one set of estimates for particulate standards based on policies now in effect. Three alternatives were considered for the control of sulfur dioxide emissions by public utilities, the greatest contributors by far to this type of air pollution. It was necessary to follow this rather cumbersome approach because the final policy on sulfur dioxide emissions has yet to be determined. The first alternative—physical coal-cleaning—would not permit public utilities to meet these standards. The second alternative, which would combine deep-cleaning with some Myers-TRW chemical coal-cleaning would allow public utilities to meet primary, but not secondary, standards. The third alternative—a combination of chemical coal-cleaning and sulfur oxide flue-gas scrubbing—would permit all establishments in the state to meet both primary and secondary standards.

The total capital costs of particulate abatement plus Alternative 1 would be about \$353 million, with a rise to \$439 million under Alternative 2. There would be a substantial jump to \$976 million under Alternative 3. Annual operating costs under the first alternative would be slightly more than \$55 million, and these would increase to about \$78 million if Alternative 2 were adopted. Again, there would be a substantial jump to \$152 million under Alternative 3. These costs, as is true of all costs in the present study, are expressed in 1975 prices.

The results given above show the direct costs of the three alternatives considered. The total costs—direct and indirect—of producing a constant volume of goods and services for final sales in a given year are more meaningful.

To meet the conditions postulated under Alternative 1, the value of total interindustry transactions would increase by \$144 million or 2.8 percent. Interindustry transactions do not take into account the cost of additional imports or other final payments. These are, however, included in the change in total gross output resulting from pollution abatement. Under Alternative 1, the change in total gross output minus a constant level of production for sale to final demand would be \$323 million, in 1975 prices, or an increase of 1.6 percent.

Under the second alternative the cost of interindustry transactions would go up by \$188 million, which represents an increase of 3.7 percent. The corresponding increase in total gross output would be \$413 million or 2 percent. Finally, under the third alternative, interindustry transactions would go up by \$497 million or 9.7 percent. The change in total gross output less a constant level of final demand would come to \$864 million, and this would represent an increase of 4.3 percent.

The latter figure is our best estimate of the total cost of meeting the West Virginia primary and secondary standards resulting from the Clean Air Act of 1970. The absolute cost of clean air is enormous, but the relative increase of 4.3 percent in the "real" cost of clean air does not seem to be an unreasonable price to pay for the wide range of benefits that would be engendered by full compliance with established standards.

This study was completed before the oil embargo was imposed following the October 1973 outbreak of hostilities in the Middle East. Under the stringencies of fuel shortages there was some relaxation of SO_2 secondary standards. Present site emission standards may not be altered, but the time table for compliance may be changed. It does not seem likely at present that the secondary standards will be enforced by 1977. Thus the conclusions derived from the first part of this study may be of less immediate interest than they might have been in the absence of the recent "energy crisis."[4]

Chapter 6, which deals with the gasification of coal, may be more timely than the earlier chapters. Coal gasification could make an important contribution to the achievement of clean air standards. It is likely to make an equally important contribution to the more recent policy proposal of attempting to achieve national energy self-sufficiency by the end of the present decade.

Since techniques for converting coal to clean gas on a commercial scale have not been perfected as yet, it was necessary to simulate the economic impact of coal gasification on the West Virginia economy.[5] Two processes for converting coal to gas were simulated as typical of a broader range of possible processes. One is the atmospheric fixed-bed process now being developed by the Bureau of Mines. The second is a two-stage, fluidized-bed process proposed by West Virginia University College of Engineering. We assumed that either process could be used in conjunction with advanced power generation systems. For purposes of comparison, we also simulated the expansion of an equal amount of conventional steam-generating capacity, assuming that the conventional plant would be fitted with effective sulfur dioxide scrubber systems.

Expansion of conventional plants would have a larger total economic impact on the West Virginia economy than either of the gasification processes. We estimated that expanding conventional plant capacity by 2,500 megawatts would add about $1.6 billion to the state's total gross output, an increase of 3.5 percent.[6] By comparison, an atmospheric fixed-bed gasification complex of the same capacity would add only $1.5 billion or 3.3 percent. The dual-stage, fluidized-bed gasification complex would add even less—$1.4 billion or 3.1 percent. The employment impacts would be in reverse order, however. When all of the direct and indirect effects are taken into account, the dual-stage, fluidized-bed gasification complex would add an estimated 14,394 new jobs in the state. An atmospheric fixed-bed complex would add 14,188, while the expansion of conventional plant capacity would create only an estimated 12,127 new jobs.

If both regional development and environmental protection objectives are taken into account, the dual-stage, fluidized-bed complex clearly comes in as a winner. This conclusion is reinforced when we consider the unit cost of production of electric energy by each of the three simulated complexes. The unit cost of the incremental output of the conventional plant we estimated at 1.57¢ per kilowatt hour. This would be reduced by the atmospheric fixed-bed process to 1.43¢ per kilowatt hour. But the lowest of all, 1.38¢ per kilowatt hour, would be achieved by the dual-stage, fluidized-bed process. Thus in terms of both the cost of production and of employment impacts the fluidized-bed process ranks first.

What conclusions with respect to regional economic development can be derived from the simulation of coal gasification processes in West Virginia? First, when large sums of money are being expended for the development of coal gasification processes, simulation might provide useful guidelines for the allocation of research funds. Clearly, these allocations cannot be based entirely on economic analysis. The technological feasibility of each of the processes must also be taken into account. But if there is no reason for choosing between the two processes simulated in this study on technological grounds, a strong case can be made for the fluidized-bed process on economic grounds. It would generate more new employment, and the end product—electricity—would be produced at lower unit costs.

The second conclusion is perhaps even more directly related to regional development policy. Regional development policy in the United States has been concerned almost entirely with direct assistance to private establishments locating or expanding in development areas, and investment in public facilities in those areas. Public facility investment ranges from the improvement of community amenities to new highway systems. A supplement to these approaches, however, could be direct investment in new processes—such as coal gasification—even though these could operate successfully only in selected development areas.

Low-BTU goal gasification in West Virginia, or in other parts of Appalachia with abundant coal supplies, offers an excellent illustration of the type of public investment that can achieve multiple policy objectives. A successful coal gasification program would contribute to the objectives of the Clean Air Act of 1970, and at the same time it would stimulate the economic development of an area that has been a prime target of federal regional development programs. Furthermore, the successful development of a low-BTU coal gasification process could make a significant addition to the nation's electric energy generation capacity. This could, in turn, permit petroleum to be diverted to other uses. Thus a successful coal gasification program would contribute to a third policy objective, that of increasing the degree of national energy self-sufficiency.

Notes

Notes

Chapter 1
Introduction

1. The interested reader will find this legislative history in Sar A. Levitan, *Federal Aid to Depressed Areas*, Baltimore: The Johns Hopkins Press (1964), pp. 1-29.

2. For further details about EDA see Niles M. Hansen, *Rural Poverty and the Urban Crisis*, Bloomington, Indiana: Indiana University Press (1970), pp. 138-159.

3. A detailed account of the establishment and operation of the Appalachian Regional Commission will be found in Monroe Newman, *The Political Economy of Appalachia*, Lexington, Massachusetts: Lexington Books, D.C. Heath and Company (1972).

4. Federal authority to deal with air pollution was first expressed in the Clean Air Act of 1963, which allowed the federal government to intervene in interstate pollution problems. This act was amended several times, and the 1965 amendment permitted, in principle at least, federal regulation of emissions by new motor vehicles.

5. The details of the West Virginia Implementation Plan are discussed below.

6. The model is described in W.H. Miernyk et al., *Simulating Regional Economic Development*, Lexington, Massachusetts: Lexington Books, D.C. Heath and Company (1970).

7. The sectoral details of this analysis are not presented in this report, but complete details are given in William H. Miernyk, "Environmental Management and Regional Economic Development: The Case of Appalachia," prepared for *Environmental Management and Regionalism*, Essays in honor or Morris E. Garnsey, Charles W. Howe (ed.), forthcoming.

8. For a discussion of these models and their limitations see *Guidelines for the Development of Air Quality Standards and Implementation Plans*, U.S. Department of Health, Education and Welfare (May 1966), p. 40.

9. The coefficients for Sector 43 (trucking and warehousing) include estimates of the pollutants emitted by moving trucks. But the *reduction* in emissions (which are very small) apply only to stationary sources in this sector.

Chapter 2
Method of Analysis

1. William H. Miernyk et al., *Simulating Regional Economic Development*, Lexington, Massachusetts: Lexington Books, D.C. Heath and Company (1970).

2. John H. Cumberland, "A Regional Inter-Industry Model for Analysis of Development Objectives," *Papers*, The Regional Science Association, Vol. XVII (1966), pp. 65-94.

3. For the California study see B.M. Lofting and P.H. McGauhey, *Economic Evaluation of Water*: Part II, *An Inter-Industry Analysis of the California Water Economy*, Contribution No. 67, Water Resources Center, Berkeley: University of California (1963), pp. 68-73. The complete report of the *Colorado River Basin Study* was not published, but there was a limited distribution of a highly detailed mimeographed version. See B. Udis et al., *Colorado River Basin Study*, Boulder, Colorado: Bureau of Economic Research, University of Colorado (1968). For a concise summary of the methodological aspects of the *Colorado River Basin Study* see William H. Miernyk, "An Inter-Industry Forecasting Model with Water Quantity and Quality Constraints," *Systems Analysis for Great Lakes Water Resources, Proceedings of the Fourth Symposium on Water Research of the Ohio State University*, Columbus: Water Resources Center (October 1969), pp. 29-58.

4. See Wassily Leontief, "Environmental Repercussions and the Economic Structure: An Input-Output Approach," *The Review of Economics and Statistics*, Vol. LII (August 1970), pp. 262-271. Wassily Leontief and Daniel Ford, "Air Pollution and the Economic Structure: Empirical Results of Input-Output Computations," in A. Brody and A.P. Carter (eds.), *Input-Output Techniques*, New York and Amsterdam: North-Holland Publishing Company, American Elsevier (1972), pp. 9-30.

5. Leslie Ayres, Ivars Gutmanis and Adele Shapanka, *Environmental Implications of Technological and Economic Change for the United States, 1967-2000: An Input-Output Analysis*, Washington: International Research and Technology Corporation (1970).

6. Robert E. Kohn, "Input-Output Analysis and Air Pollution Control," Paper presented to the Universities—National Bureau Committee for Economic Research Conference on Economics of the Environment (November 1972).

7. In some cases, however, there may be more disaggregation at the state level than in national tables. In West Virginia, for example, the basic (1965) transactions table had separate deep and surface coal mining sectors. In the national table these are lumped together as a single sector.

8. The reason for the distinction at the regional level is that regional coefficients depend upon both current technology and trade patterns, whereas national coefficients are determined almost entirely by technical considerations.

9. Further details on the construction and interpretation of static, open models are given in William H. Miernyk, *The Elements of Input-Output Analysis*, New York: Random House (1965).

10. The discussion in the text draws heavily on William H. Miernyk, "Regional Input-Output Pollution Abatement Model," *IEEE Transactions on Systems, Man and Cybernetics* (November 1973).

11. For a discussion of this method of projection, see Miernyk, *The Elements of Input-Output Analysis*, pp. 117-125.

12. The unconstrained projections are given in Miernyk et al., op. cit.

13. Leontief and Ford, op. cit., p. 22.

14. Ibid., pp. 12-14.

15. The direct and indirect coefficients were calculated by multiplying the vector of particulate emissions by the West Virginia Leontief inverse matrix for 1965.

16. For the national counterparts of these coefficients see Leontief and Ford, op. cit., pp. 15-17.

17. Similarly, there are cases where hydrocarbons are emitted with sulfur dioxide and particulates, and in many cases the reduction of particulate emissions leads to a reduction in hydrocarbon emissions.

18. Leontief and Ford, op. cit., p. 12.

19. Data were not collected for emissions of nitrogen oxides in West Virginia. We are aware that nitrogen oxides may be harmful to plant life, including some commercial operations such as apple and peach orchards that are quite important in some of the eastern West Virginia counties. Information on emission standards and the control of stationary sources was still being developed at the time data were collected for this study, however, and we were not willing to use the scanty and preliminary information that was available.

Chapter 3
The Control of Particulate Emissions

1. If a company's pollution control strategy includes a shift from relatively high- to relatively low-sulfur coal, this decision will have an effect on the size and type of electrostatic precipitator to be installed. The efficiency of a precipitator will usually be affected by the sulfur content of the coal burned, since the electrical resistance of the ash is influenced by sulfur content. In general, there is an *inverse* relationship between efficiency and sulfur content. Thus, as sulfur content declines, larger precipitators must be installed to achieve a given level of net dust removal.

2. We do not pretend that the visited establishments constitute a random sample of all establishments in each sector. A few firms contacted were unwilling to cooperate, and others failed to respond to mail inquiries or attempts to contact responsible officials by telephone. Firms refusing to cooperate represented a small percentage of the total number contacted, however.

3. These estimates were provided by company officials at WVAPCC hearings. They were supplemented by data obtained from various engineering publications including: *Chemical Week, Journal of Air Pollution Control, Environmental Science and Technology, Mining Congress Journal, Minerals Processing,* and *Iron and Steel Engineer.*

4. See, for example, D. LeSourd et al., *Comprehensive Study of Specific Air Pollution Sources to Assess the Economic Effects of Air Quality Standards*, Research Triangle Institute, for the Environmental Protection Agency (1970), and *Air Pollution Engineering Manual*, 909 AP 40, NAPCA (1969). These studies did not adequately consider the cost of retrofitting in existing plants. This conclusion is corroborated by the costs cited in a detailed study by the Midwest Research Institute, *Particulate Pollutant System Study* (1971). The authors of the MRI study based their estimates on interviews with company officials. Their estimates are closer to ours than to the studies cited above.

5. A private organization, the Economics Priorities Council, has focused on prominent corporations in the electric power and steel industries. In their report on the American Electric Power Company, for example, they concluded that particulate abatement equipment could be installed in the 1100 MW Philip Sporn plant for $3.5 million. They did not, however, take into account the costs of retrofitting and the higher costs involved in the shift to low-sulfur fuel. The most recent AEP estimates of the cost of meeting particulate standards in this plant range between $30 and $55 million.

6. This is not a significant problem, however, since the separate cost estimates of Chapters 3 and 4 are combined in Chapter 5.

7. The cost estimates were obtained for the years shown in Table 3-1. In order to adjust the coefficients of the West Virginia input-output model on the basis of these cost increases, it was first necessary to *deflate* all of the cost estimates to 1965 dollars. The West Virginia model does not allow for price adjustments so all projections to 1975 were originally expressed in 1965 dollars. These were then inflated, using data from the sources indicated in Table 3-1, to obtain the results that are expressed in 1975 dollars.

8. Estimates of capital expenditures by electric utilities were revised upwards at WVAPCC hearings held after this study was completed. The more recent estimates reflected the added costs of retrofitting and of larger precipitators required by shifts to low-sulfur coal. This reinforces our view that the capital cost estimates in this study are on the conservative side.

9. See Miernyk et al., *Simulating Regional Economic Development*, Lexington, Massachusetts: Lexington Books, D.C. Heath and Company (1970), pp. 280-283.

10. The capital costs would, of course, be amortized over a period of years.

Chapter 4
The Control of Sulfur Dioxide Emissions

1. This was written before final Congressional action on the Emergency Energy Bill introduced in 1973.

2. A recent change in the regulation listing a maximum of 45,000 lbs. of

SO_2/hr. from a single plant, rather than the earlier 32,000 lbs. SO_2/hr., allows the American Electric Power System's new John E. Amos plant to burn low-sulfur West Virginia coal without further action as previously anticipated. This will make it easier for the same company's Philip Sporn plant to obtain low-sulfur coal. Some coal may have to be obtained from other mines in the same area, however, to reduce the average sulfur content to acceptable levels. The added cost of redistribution of in-state coal supplies is difficult to estimate and has not been considered in the present study.

3. Capital spending at the rate of $96 per kilowatt of capacity assumes that all of the abatement equipment will be installed in existing plants and reflects the high cost of retrofitting. If the abatement equipment is installed in new plants at the time of construction or if retrofitting costs are minimal, capital spending is assumed to be at the rate of $48 per kilowatt of capacity. The costs involved when the lower end of the capital spending range is assumed are given in Appendix Table 4A-2.

4. WVAPCC *Hearings*, November 1972-August 1973.

5. *Steam Electric Power Factors*, 1972 edition, Washington, D.C.: National Coal Association; Private Communication, 1973, R.R. Bertrend, Exxon Oil Company; Private Communication, 1973, Ed Wiley, W. Va. Coal Association; *Mineral Industries Yearbook*, U.S. Department of Interior, 1963-1972.

6. C. Davis, Babcock and Wilcox, Private Communication, 1973.

7. J.K. Burchard et al., *Some General Economic Considerations of Flue-Gas Scrubbing for Utilities*, Environmental Protection Agency (October 1972); S. Katell, "Alkalized Alumina Process," *CEP*, 62 (60) (1966); Hitman Associates, Inc., *Cost Nomographs of Selected Sulfur Dioxide Abatement Methods* (1972).

8. WVAPCC Hearing, 1972-1973; D.C. Gifford, "Will County Unit 1–Limestone Wet Scrubber," Conference Proceedings on Sulfur in Utility Fuels: The Growing Dilemma, *Electrical World*, McGraw-Hill (1972). As noted earlier, this is the maximum rate, based on the assumption that all abatement equipment is to be retrofitted.

9. Mitre Corporation, *The Physical Desulfurization of Coal–Major Considerations of SO_2 Emission Control*, PB210-273 (1972); Private Communication, 1972, R.A. Myers, TRW, Inc.

10. It is possible that some utilities could meet secondary standards by fuel-switching, but because of limited supplies of low-sulfur fuel in this region, fuel-switching cannot be considered to be an effective long-term solution. Some utilities that have already switched to low-sulphur coal will have to turn to flue-gas scrubbing in the future. See WVAPCC *Hearings* (August 14, 1973).

Chapter 6
Simulating Low BTU Coal Gasification

1. We did not rule out the possibility, however, that the decline in West Virginia coal production after 1965 might have been due *in part* to the

imposition of local air quality standards in some large eastern metropolitan areas.

2. Chauncey Starr, "Energy and Power," *Scientific American*, Vol. 225 (Sept. 1971), p. 37.

3. *Summary Report of the Cornell Workshop on Energy and the Environment*, Committee on Interior and Insular Affairs, U.S. Senate, A National Fuels and Energy Policy Study, Serial No. 92-33 (May 1972), p. 43.

4. In Appalachia as a whole there are over 255 billion tons of known reserves of deep coal. See Council on Environmental Quality, *Coal Surface Mining and Reclamation*, Committee on Interior and Insular Affairs, U.S. Senate, Serial No. 93-8 (92-43), p. 52.

5. Ralph E. Lapp, "We're Running Out of Gas," *The New York Times Magazine*, (March 19, 1972), p. 35.

6. These estimates have been corroborated by McKetta. See John J. McKetta, "The Energy Crisis: On and On and On . . . ," *Chemical Engineering Progress* (August 1973), p. 51.

7. *The Encyclopedia Britannica*, Vol. XI, No. 11-12, p. 483.

8. For a more detailed and technical description of this process, see H.C. Hottel and J.D. Howard, *New Energy Technology: Some Facts and Assessments*, Cambridge, Massachusetts: The MIT Press (1971), p. 110-117. A lucid, non-technical description of coal gasification is given by A.M. Squires in "Clean Power from Dirty Fuels," *Scientific American*, Vol. 227 (October 1972), pp. 26-35.

9. Hottel and Howard, op. cit., pp. 144-145.

10. On the development of gas turbines, see Squires, op. cit., p. 27. Turbines, which would operate at the required temperatures, have not been fully developed as yet, but engineers and metallurgists appear to be optimistic about an early solution to the temperature problem.

11. For a general discussion of various low-BTU gasification processes, see Hottel and Howard, op. cit., pp. 144-161, and Squires, op. cit.

12. See, for example, S. Katell, P.S. Lewis and P. Wellman, "The Economics of Producer Gas at Atmospheric and Elevated Pressures," 165th Annual Meeting of the American Chemical Society (April 1973); C.Y. Wen, R.C. Bailie, C.W. Lin, and W.S. O'Brien, "Production of Low-BTU Gas Involving Coal Pyrolysis and Gasification," 165th Annual Meeting of the American Chemical Society (April 1973); and C.Y. Wen, *Production of Electricity via Coal and Coal-Char Gasification*, Office of Coal Research Report (June 1973).

13. A pressurized fixed-bed process (also being developed by the Bureau of Mines) was analyzed and simulated. In terms of economic impact, however, there is little difference between the two fixed-bed processes. Details are given here only for the atmospheric fixed-bed process.

14. We estimate that the thermal efficiency of the fluidized-bed system is at least 2 percent greater than that of the single-stage system.

15. Personal communication from Arthur M. Squires, dated November 19,

1973. Because of the low velocity of the simulated system it would require a larger gasifier than the one proposed by Squires, or several gasifiers of the same size, to produce a comparable volume of gas.

16. J.K. Burchard, G.T. Rochell, W.R. Schoffield, and J.O. Smith, *Some General and Economic Considerations of Flue-Gas Scrubbing for Utilities*, Environmental Protection Agency (October 1972).

17. Down time for repairs at each plant was assumed to average five weeks per year. This might seem to be a high rate of capacity utilization. But we assumed that the new plants would be operated as close to peak capacity as possible; any slack in the demand for energy would lead to a reduction in capacity utilization in older units in the power system.

18. To adjust costs at different levels of capacity than those used by Wen, Buchard, Katell, and their associates, we used the relation:

New Cost = Old Cost (New Capacity/Old Capacity)n, where n = the scale factor.

Scale factors used were 1.0 for gasifiers and power equipment, 0.8 for facilities, and 0.6 for the ammonium sulfate plant.

19. See Miernyk et al., *Simulating Regional Economic Development*, Lexington, Massachusetts: Lexington Books, D.C. Heath and Company (1970), p. 279.

20. Ibid.

21. Ibid., pp. 58-59.

22. Ibid., pp. 126-186.

23. The prototype model suggested by Wen and his associates is for a single unit of 500 MW capacity. In blowing up Wen's data to 2,500 MW complexes, we assumed that each complex would operate along a conventional Leontief production function—that is, there would be proportional increases in all inputs as output increases. If there actually should be economies of scale, our estimates would overstate labor input requirements.

24. The value of final demand for each simulated sector is approximated by the entries in row 49 of Table 6-3. The difference between the values in this row and the actual final demand figures used to make the 1975 projections in each case represents the internal consumption of electric energy produced by each of the complexes. The imputed value of this consumption was treated as an *intraindustry* transaction in each complex.

25. *Economic Report of the President*, Washington: U.S. Government Printing Office (January 1973), pp. 196 and 248.

26. At present, most state commissions are allowing a return of between 7 and 8 percent on original investment.

27. It would be quite easy to add the return on investment to this figure; but the numbers given in Column 3 are cost of production figures only, and these are what we want to compare.

28. The details of the West Virginia dynamic inverse are given in Miernyk et

al., op. cit., pp. 165-186. For a general description of the theory and operation of the dynamic inverse see Wassily Leontief, "The Dynamic Inverse" in A.P. Carter and A. Brody, *Contributions to Input-Output Analysis*, New York and Amsterdam: North-Holland/American Elsevier (1970), pp. 17-46.

29. Squires, op. cit., p. 35.

Chapter 7
Environmental Management and Regional Economic Development

1. The major declines in total output in primary metals and chemicals, and the large reduction in investment in the state as a whole, are due to the reverse accelerator effects generated by the West Virginia dynamic model.

2. Sectors 2 and 3, underground and strip coal mining, respectively, were combined for the purposes of this study so the tables in the text show changes in only 16 sectors.

3. We did not go into the issue of price changes induced by the cost of increases. These have been investigated by two of the graduate research assistants who worked on the project, however. See Frank A. Giarratani and Luther C. Thompson, "Air Pollution Abatement, Technological Change, and Relative Prices: A Regional Input-Output Analysis," Morgantown, West Virginia: Regional Research Institute, MR-15 (1974); and Frank Giarratani, "Air Pollution Abatement: Output and Relative Price Effects: A Regional Input-Output Simulation," Regional Research Institute, MR-16 (1974).

4. Any delay in the implementation of air quality standards will most likely lead to higher costs of abatement than those given in this study because of the effects of continued inflation.

5. Work is now under way on processes to convert coal to both high-BTU and low-BTU gas. Our study focused entirely on low-BTU gasification since this type would be both technically and economically feasible in West Virginia.

6. An observant reader will notice that while the absolute change in TGO is much larger than the increase resulting from pollution abatement expenditures, the *relative* increase is smaller. This is because different bases were used to compute the percentage changes. In the case of abatement expenditures, we measured the change in TGO *minus* a constant vector of final demand. When simulating the effects of coal gasification, however, we not only included final demand when calculating changes in TGO, but also assumed substantial increases in total final sales due to inclusion of the simulated sectors. The objective was different in the two cases. In the first, we estimated the "real" cost of pollution abatement; in the second, we estimated the economic impacts of the *new* activities simulated.

Selected Annotated Bibliography

Selected Annotated Bibliography

Frank A. Giarratani

General

Fisher, A.C., Krutilla, J.V., Cichetti, C.J., "The Economics of Environmental Preservation: A Theoretical and Empirical Analysis," *American Economic Review*, September 1972, Vol. LXII, Number 4, pp. 605-19.
The paper is an extension of earlier work done by Krutilla on the preservation of natural environments. A model is developed for the allocation of natural environments between preservation and development.

Fisher, J.L., "An Introduction to Environmental Economics: Population and Natural Resources" in *The Economics of Environmental Problems*, F.C. Emerson (Ed.), Division of Research, Graduate School of Business Administration, University of Michigan, 1973.
A discussion of the trends in population, natural resources, and the environment, with policy suggestions for some resource problems.

Fisher, J.L., "Impact of Population on Resources and the Environment," *American Economic Review, Papers and Proceedings*, May 1971, Vol. 61, Number 2, pp. 393-98.
The purpose of this paper is to examine the principal impacts of population growth and distribution in the United States on resource availability and on environmental quality.

Kneese, A.V., "Background for the Economic Analysis of Environmental Pollution," *Swedish Journal of Economics*, March 1971, Vol. 73, Number 1, pp. 1-24.
Provides background information on technological aspects of environmental problems.

Kneese, A.V., "Environmental Pollution: Economics and Policy," *American Economic Review, Papers and Proceedings*, May 1971, Vol. LXI, Number 2, pp. 153-66.
A discussion of some models concerning common property resources (air and water) and their policy implications. (See also articles on externalities.)

Lave, L.B., "The Economic Costs of Air Pollution" in *The Economics of Environmental Problems*, F.C. Emerson, (ed.) Division of Research, Graduate School of Business Administration, University of Michigan, 1973.
This paper outlines a cost-benefit approach to evaluating air pollution abatement. Some difficulties of this approach are discussed generally and specific attention is given to estimates of direct costs and benefits. *Cost estimates:* From "The Economics of Clear Air," Report to the Congress of the United States. *Benefit estimates:* Reviews and uses several different

sources including the SO_2 criteria documents recently cast in doubt, and briefly outlines some primary attempts at statistical analysis of the air pollution/health relationship.

U.S. Government Printing Office, Washington, D.C., *Some Environmental Implications of National Fuels Policies*, prepared for the Committee on Public Works, United States Senate, December 1970.
Contains projections of energy demands and fuel requirements as well as a discussion of the various pollutants associated with fossil fuels and the available control technology.

U.S. Government Printing Office, Washington, D.C., *Clean Air For the Nation: The Report of the President's Task Force on Air Pollution*, August 1970.
Summary of the current information on control techniques and biological effects of air pollution.

U.S. Government Printing Office, Washington, D.C., *Environmental Quality: The First Annual Report of the Council on Environmental Quality*, August 1970.
CEQ's annual report on the status of federal and local pollution control efforts.

U.S. Government Printing Office, Washington, D.C., *Environmental Quality: The Second Annual Report of the Council on Environmental Quality*, August 1971.
CEQ's annual report on the status of federal and local pollution control efforts.

U.S. Government Printing Office, Washington, D.C., *Environmental Quality: The Third Annual Report of the Council on Environmental Quality*, August 1972.
CEQ's annual report on the status of federal and local pollution control efforts.

U.S. Government Printing Office, Washington, D.C., *A Review of Energy Policy Activities of the 92 D Congress, First Session*, prepared for the Committee on Interior and Insular Affairs, United States Senate, 1972.
A comprehensive review of issues concerning the nations energy needs and their implications. Each major energy source is considered in turn.

Input-Output Applications

Ayres, L., Gutmanis, I. and Shapanka, Adele. "Effects of Technological Change on, and Environmental Implications of an Input-Output Analysis for the United States, 1967-2000" in Part II of *Environmental Implications of Technological and Economic Change for the United States, 1967-2000: An Input-Output Analysis*, prepared for Resources for the Future by International Research and Technology Corp., Washington, D.C., 1970.
This paper uses an input-output model to determine wastes from selected industries in most sectors of the economy and to assess the cost of abating these wastes, through to the year 2000.

Cumberland, John H., "A Regional Interindustry Model for Analysis of Development Objectives," *Papers and Proceedings of the Regional Science Association*, Vol. 17, 1966, pp. 65-94.

Cumberland included environmental quality among his criteria for the analysis of economic development programs. This paper is the pioneering effort to account for environmental factors in an input-output framework. He suggests the introduction of an environmental balance row and column in the conventional regional interindustry model.

Lavrent, E.A., and Hite, J.C., "Economic Ecologic Linkages and Regional Growth: A Case Study," *Land Economics*, February 1972, Vol. XLVIII, Number 1, pp. 70-72.

This paper uses the Leontief static inverse and a matrix showing inflow from the environment and the residual outflow to the environment associated with one dollar's worth of gross output of each sector in the model. The environmental impact of projected growth in each sector is evaluated with the aid of "environmental income multipliers," i.e., environmental repercussions of a one-dollar increase in pecuniary income generated from a given sector. (See also articles on input-output application to environmental problems.)

Leontief, W., "Environmental Repercussions and the Economic Structure: An Input-Output Approach," *Review of Economics and Statistics*, August 1970, Vol. 52, Number 3, pp. 262-71.

Links undesirable by-products and valuable but unpaid-for inputs directly to physical relationships of our economic system. Generally, Leontief shows how such externalities can be incorporated into the conventional input-output framework.

Leontief, W., and Ford, D., "Air Pollution and the Economic Structure: Empirical Results of Input-Output Computations," in *Input-Output Techniques*, edited by A. Brody and A.P. Carter, North-Holland Publishing Company, 1972, pp. 9-30.

This article is the empirical implementation of the theoretical foundation laid by Leontief in his 1970 article incorporating environmental "externalities" in an Input-Output framework. A matrix of direct pollution output coefficients and a conventional matrix of interindustry coefficients are used to analyze the dependence of five types of air pollution on the input-output structure of the American economy. (See also W. Leontief, "Environmental Repercussions and the Economic Structure: An Input-Output Approach," *Review of Economics and Statistics*, August 1970, Vol. 52, Number 3, pp. 262-71.)

National Planning Association, *A Projection Simulation Model for Analyzing Environmental Impacts of U.S. Economic Growth and Technical Change*, a Pre-proposal discussion draft, August 23, 1972.

A survey of existing input-output model applications to environmental impact analysis and discussion of the NPA's proposed model with some emphasis on dealing with technological change.

Regional Models

Lakshmanan, et al., *The OAP Regional Economic Model Utilization-Phase 1, Volume I: Simulation and Analysis*, a report by Causal Research Company for EPA, January 7, 1972.
Application of Keynesian-type macro-model to the analysis of regional economic impact of air pollution abatement. (See also Lakshmanan's article in *Environmental and Planning*, 1972, Vol. 4, pp. 73-97.)

Lakshmanan, T.R., and Fu-cheu Lo, "A Regional Economic Model for the Assessment of Effects of Air Pollution Abatement," *Environment and Planning*, 1972, Vol. 4, pp. 73-97.
The model is a cross-sectional Keynesian-type regional macro-model that is connected to a national input-output model via a regional share (location quotient) matrix. Investigates the relative economic effects of three strategies reflecting control costs based on the Clean Air Act of 1967.

Russell, C.S., "Application of Micro Economic Models to Regional Environmental Quality Management," *American Economic Review, Papers and Proceedings*, May 1973, Vol. LXIII, Number 2, pp. 236-43.
The paper begins with a review of the major types of models available and then discusses the present model being developed by Resources for the Future. The model is designed to optimize subject to ambient quality standards and deals with air, water, and solid waste problems simultaneously within a linear programming framework.

Optimizing Models

Keeler, E., Spence, M. and Zeckhauser, R., "The Optimal Control of Pollution," *Journal Economic Theory*, February 1972, 4(1) pp. 19-34.
The paper examines optimal allocation of income to consumption, capital formation, and pollution in models where pollution is entered in the utility function. It also discusses dynamic aspects of pollution control on convergence to steady-state expenditure patterns. (See also articles on optimal control theory and steady state growth.)

Kohn, R.E., "Application of Linear Programming to a Controversy on Air Pollution Control," *Management Science*, Vol. 17, No. 10, June 1971.
Uses linear programming techniques to evaluate certain charges that it is economically infeasible for purposes of air pollution control to impose restrictions on the sulfur content of coal. He finds that such regulation is justified in certain types of stokers but not in others.

Kohn, R.E., "Joint-Outputs of Land and Water Wastes in a Linear Programming Model for Air Pollution Control," *Social Statistics Section Proceedings of the American Statistical Association*, 1970.

Considers three external waste stream created by air pollution abatement measures (liquid wastes, landfill wastes, and thermal discharge to rivers) in a linear programming model to account for their impact on the costs of air pollution abatement. (See also Agnes and Kneese, "Production, Consumption, and Externalities," *American Economic Review*, June 1969, Number 3, pp. 282-297.)

Kohn, R.E., "Linear Programming Model for Air Pollution Control: A Pilot Study of the St. Louis Airshed," *Journal of Air Pollution Control Association*, February 1970, Vol. 20, Number 2.

Linear programming approach for finding least cost alternative to meet given standards.

Kohn, R.E., "Optimal Air Quality Standards," *Econometrica*, November 1971, Vol. 39, Number 2, pp. 983-85.

Uses linear programming techniques to develop a set of alternative air quality levels for sulfur dioxide and particulates which can be attained at the same total cost and translates this into a frontier of trade-off possibilities.

Kohn, R.E., "A Cost-Effectiveness Model for Air Pollution Control With A Single Stochastic Variable," *Journal of the American Statistical Association*, March 1972, Vol. 67, Number 337, pp. 19-22.

Cost of abatement is minimized for a given set of air quality goals and for varying degrees of confidence that the goals will be achieved. Shows that the costs of increased certainty rise rapidly and suggests that goals should be expressed not as maximum concentrations, but also the minimum probabilities that these goals will not be exceeded. (See also Kohn's numerous other articles on linear programming methods of analysis.)

Kohn, R.E., "Abatement Strategy and Air Quality Standards," in *Development of Air Quality Standards*, edited by A. Atkisson, R. Gaines, Charles E. Merril Publishing Company, 1970.

Application of linear programming to estimate least cost alternative for air pollution control. Model generally attempts to minimize cost subject to a particular control method's activity level.

Kohn, R.E., and Burlingame, D.E., "Air Quality Control Model Combining Data on Morbidity and Pollution Abatement," *Decision Sciences*, Vol. 2, July 1971, pp. 300-310.

The article develops a pollution damage function which incorporates the costs and technology for pollution abatement in the St. Louis Airshed. Standards are developed which minimize morbidity subject to a given control budget constraint.

Pricing of Pollution

Frech III, H.E., "Pricing of Pollution: the Coase Theorem in the Long Run," *Bell Journal of Economics and Management Science*, Spring 1973, Vol. 4, Number 1, pp. 316-19.

This paper is a reply to Tybout's earlier paper "Pricing Pollution and Other Negative Externalities," *Bell Journal of Economics and Management Science*, Spring 1972, Vol. 3, Number 1, pp. 252-66. Frech claims to demonstrate that Tybout's result is based on a fundamental error in the analysis of costs for the polluting and polluted industries. (See also articles on the Coase Theorem.)

Kneese, A.V., "Pollution and Pricing," *American Economic Review*, December 1972, Vol. LXII, Number 5, p. 958.
A comment on Stein's article recommending federal government control of pollution regulation. Kneese argues that Stein fails to separate measures of pollution damage and measures of emissions in his analysis. (See Stein's article, *American Economic Review*, September 1971, pp. 531-37.)

Tybout, R.A., "Pricing Pollution and Other Negative Externalities," *Bell Journal of Economics and Management Science*, Spring 1972, Vol. 3, Number 1, pp. 252-66.
Equilibrium conditions are described with pollutants or "externalities of production" treated as substitute products. The results are contrasted with those based on traditionally assumed fixed proportions between product and waste outputs. (See also articles on externalities in production and market failure.)

Externalities

Ayres, R.M., and Kneese, A.V., "Production, Consumption, and Externalities," *American Economic Review*, June 1969, Vol. LIX, Number 3, pp. 282-97.
The authors argue that externalities have traditionally been viewed as exceptional cases and that at least one class of externalities, residuals, are an inherent and general part of the production and consumption process. They exhibit a formal mathematical framework for tracing residuals flows in the economy and relate it to a general equilibrium model of resource allocation. (See also articles on externalities.)

Baumal, W.J., and Oates, W.E., "The Use of Standards and Prices for Protection of the Environment," Swedish Journal of Economics, March 1971, Vol. 73, Number 1, pp. 42-54.
Application of traditional Pigouvian unit taxes to control externalities. Analysis suggests that if taxes on waste emissions, sufficient to attain given standards, are applied they can achieve a specified reduction in pollution at minimum cost. (See, for example, Oates, W.E., *Fiscal Federalism* (Harcourt Brace Jovanovich, Inc., 1972) for a good discussion of Pigouvian taxes.)

Gramm, W.P., "A Theoretical Note on the Capacity of the Market System to Abate Pollution," *Land Economics*, August 1969, Vol. XLV, Number 3, pp. 365-68.
This article treats pollution abatement as a type of product differentiation

and attempts to show that, given public awareness of the pollution problem and the public's desire to abate pollution, it is theoretically possible that the market may account for externalities of this type. (See also articles on externalities.)

Zerbe, R.O., "Theoretical Efficiency in Pollution Control," *Western Economic Journal*, December 1970, Vol. 8, No. 4, pp. 364-76.
Examines efficiency of different rules of liability, taxes, and subsidies. The argument that the polluter's liability is necessary to force him to initiate negotiations is incorrect. (See also articles on Coase Theorem.)

Policy

Goldman, M.I., "Externalities and the Race for Economic Growth in the U.S.S.R.: Will the Environment Ever Win," *Journal of Political Economy*, March-April 1972, Vol. 80, No. 2, pp. 314-27.
Though socialist governments would seem theoretically to have an advantage in dealing with environmental problems, the USSR is having serious difficulties. There is no evidence to indicate that nationalization of the means of production will provide any cure-all for environmental problems.

McNamara, R.S., *Address to the United Nations Conference on The Human Environment*, a publication of the International Bank for Reconstruction and Development, Washington, D.C., June 8, 1972.
Discusses briefly the growth-environment trade-off and stresses the necessity for continued growth if we are to deal with human problems.

Peltzman, S. and Tideman, T.N., "Local Versus National Pollution Control: Note," *American Economic Review*, December 1972, Vol. LXII, No. 5, pp. 959-63.
This note is a reply to Stein's argument in favor of federal government responsibility for pollution control. The authors argue that Stein fails on at least three counts:

1. He ignores the spatial aspects of pollution.
2. Fails to distinguish between the desirability of a single price and that of having a single price setter.
3. Fails to distinguish between the short and long run.

(See Stein's article, *American Economic Review*, September 1971, Vol. 61, pp. 531-37.)

Rothenberg, J., "The Evaluation of Alternative Public Policy Approaches to Environmental Control," in *The Economics of Environmental Problems*, F.C. Emerson (ed.), Division of Research, Graduate School of Business Administration, University of Michigan, 1973.
The article looks at pollution (particularly congestion) as a public goods problem. At some point of joint use the good in question loses some of its

"publicness," referred to as inter-user interference. The author then discusses some policy suggestions, stressing optimization of pollution as opposed to minimization, that is the trade-off between pollution control and the production of other goods. An argument is also made that from a policy viewpoint controlling congestion (where all lose equally) is easier than dealing with pollution (where some benefit, some lose). (See also articles on market failures.)

Stein, J.L., "The 1971 Report of the President's Council of Economic Advisors: Micro-Economic Aspects of Public Policy," *American Economic Review*, September 1971, Vol. 61, pp. 531-37.

Stein discusses methods of achieving environmental quality goals: charges for emissions of pollution, and transferable environmental usage certificates. He concludes that the latter would present consumers with the realization of the trade-off between output and the environment. He also argues that responsibility for pollution regulation should rest primarily with federal, not state and local governments because local responsibility would mean differences in charges between localities, violating necessary conditions for Pareto optimality in production and exchange. This position is opposed to that of the President's Council of Economic Advisors.

Miscellaneous

Evans, M.K., "A Forecasting Model Applied to Pollution Control Costs," *American Economic Review, Papers and Proceedings*, May 1973, Vol. LXIII, No. 2, pp. 244-52.

This paper combines the Chase Econometrics Macroeconomic Model and the Interindustry Forecasting System to estimate the effects which cost increases in specified industries will have on the economy. Direct cost estimates by industries are obtained from EPA data and these are translated into direct and indirect price increases. Estimates of effects on final demand and export price changes are made and equilibrium changes in prices and other variables are forecast by solving the complete Macro model.

Heller, W.H., "Economic Growth and Environmental Quality: Collision or Co-Existence?" A module of the General Learning Press, 1973.

This paper is directed towards the "Growth-No Growth" controversy. Heller discusses the nature of the trade-offs involved in dealing with the problem of pollution. He puts the role of economic analysis in perspective, contrasting the means and ends of economics and ecologists. (See also articles on the limits to growth.)

Jackson, R., "Zero Pollution: A Tradeoff Analysis," *Southern Economic Journal*, July 1971, Vol. 38(1), pp. 97-100.

Discusses trade-off between elimination of damage imposed on others by pollution and costs of modifying the production process as well as loss in possibilities for welfare gain from abatement.

Kohn, R.E., "Air Quality, The Cost of Capital, and the Multi-Product Production Function," *Southern Economic Journal*, October 1971, Vol. XXXVIII, No. 2, pp. 156-60.

Considers air pollution abatement in context of the theory of multi-product production functions. The final outputs are ambient air concentrations of the various pollutants. (See Hicks, J.R., *Value and Capital*, Oxford University Press, London, for a discussion of multi-product firms.)

Squires, A.M., "Clean Power From Dirty Fuels," *Scientific American*, October 1972, Vol. 227, No. 4, pp. 26-35.

A survey of coal gasification techniques in easily understandable text with diagrams.

Selected Annotated Bibliography

John T. Sears

Pollutant Emissions and Regulations

Bartok, W., Crawford, A.R., Cunningham, A.R., Hall, H.J., Manny, E.H. and Skopp, A., *Systems Study of Nitrogen Oxide Control Methods for Stationary Sources*, Vols. I and II, Esso Research and Engineering Company, for the Environmental Protection Agency, November 1969), p. 789.
Contains a summary of nitrogen oxide emissions and projected means of reducing emissions at stationary sources.

Engdahl, R.B., "A Critical Review of Regulations for the Control of Particulate Emissions," *Journal of Air Pollution Control Association*, May 1973, Vol. 23, No. 5, pp. 364-75.
Presents the history of sulfur dioxide control and air quality standards. Provides a comprehensive summary of maximum sulfur content in fuel by state and generalized statements on the status of SO_2 control processes.

Hemeon, W.C.L., "A Critical Review of Regulations for the Control of Particulate Emissions," *Journal of Air Pollution Control Association*, May 1973, Vol. 23, No. 5, pp. 376-87.
This review discusses the rationale behind the limitations on particulate emissions. It presents in general terms the limitations adopted by state and federal agencies. The strengths and weaknesses of the regulations in protecting public health are discussed.

Vandegrift, A.E., Shannon, L.J., Sallee, E.E., Gorman, P.G. and Park, W.R., "Particulate Air Pollution in the United States," June 1971, *Journal of Air Pollution Control Association*, Vol. 21, No. 6, pp. 321-28.
A comprehensive summary of present particulate emissions by industry. A more detailed account of present emissions and projected emissions is presented by the same authors in *Particulate Pollutant System Study*, Vols. I, II, and III, Midwest Research Institute, for the Environmental Protection Agency, August 1971, p. 1309.

West Virginia Air Pollution Control Commission Implementation Plan, West Virginia Air Pollution Control Commission, February 1972.
This report contains a regional breakdown of pollutant emissions and ambient pollutant levels and the regulations controlling emission. Each state has a similar plan. Further details on emissions may be found for individual states in state reports such as *Kanawha Valley Air Pollution Study*, National Air Pollution Control Administration and West Virginia Air Pollution Control Commission, March 1970.

Costs and Impact of Pollution Control on Industry

Bender, R.J., "Air Pollution Control: Its Impact on the Metal Industries," *Power*, April 1972, pp. 56-60.

Review of industry attitudes toward the multiplicity of federal agencies involved and delays in approving pollution control actions based on extensive interviews.

Economic Impact of Air Pollution Controls on Gray Iron Foundry Industry, Environmental Protection Agency, Report No. AP-74, 1970, p. 123.

A detailed analysis of techniques and costs of controlling dust from foundry operations in the United States. Includes predicted effects on prices and profitability of small operations. Typical of many such government studies. A different viewpoint on the effect of air pollution is given by Gutlow, B.S., "Impact of Foundry Pollution Control," *Environmental Science and Technology*, 1972, Vol. 6, pp. 790-94.

M.W. Kellogg Company, *Applicability of SO_2-Control Processes to Power Plants*, report to Environmental Protection Agency, EPA-R2-72-100, November 1972, p. 67.

Indicates the percentage of power plants that could install SO_2 control scrubbing systems at various cost levels.

LeSourd, D.A., Fogel, M.E., Schleicher, A.R., Bingham, T.E., Gerstle, R.W., Hill, E.L. and Ayer, F.A., *Comprehensive Study of Specified Air Pollution Sources to Assess the Economic Affects of Air Quality Standards*, Environmental Protection Agency, Report No. PB-197-647, 1970.

The first major study to assess the economic impact of air pollution control. More recent studies show it underestimated costs.

Price, F., Davidson, R. and Ross, S. (eds.), *McGraw-Hill's 1972 Report on Business and the Environment*, New York: McGraw-Hill Publications Company, 1972.

A compendium of articles on general efforts by industry to meet anti-pollution codes. Includes some articles that simply explain devices to control pollution and others citing costs of particular industry efforts.

Proceedings, How to Get Your Money's Worth, Pollution Control–'73, published by *Chemical Engineering*, New York: McGraw-Hill, 1973.

A symposium on procedures being utilized by the Chemical Processing Industry to comply with water and air pollution regulations. Articles on agency jurisdiction, techniques, and design of pollution equipment and processes are presented.

Proceedings, Sulfur in Utility Fuels: The Growing Dilemma, published by *Electrical World*, New York: McGraw-Hill, October 1972.

Proceedings of a conference on problems and approaches to control of sulfur dioxide in power plant operations. Participants included industry and government representations associated with control equipment systems and strategies.

Schwartz, I., "The High Cost of Pollution Control," *Chemical Week*, May 24, 1972, pp. 59-68; idem, May 23, 1973, pp. 49-51.
Costs of air and water pollution control efforts are broken down by year for selected companies in the chemical process industry.

Sulfur Oxide Control Technology Assessment Panel, *Projected Utilization of Stack Gas Cleaning Systems by Steam-Electric Plants*, final report submitted to Federal Interagency Committee for Evaluation of State Air Implementation Plans, April 1973, p. 93.
Includes considerations of availability of technology to install systems, effects on power-grid supplies and makes recommendations about the rate of implementation of systems in power plants to the 1980s.

Methods of Controlling Pollutant Emissions

Air Pollution Engineering Manual, U.S. Department of Health, Education and Welfare, Publication 999-AP-40, 1967.
Detailed list of processes emitting pollutants and general methods of design for control.

Allen, J.E., "Economics in Scrubber System Selection," *Iron and Steel Engineer*, June 1972, pp. 72-74.
Non-technical discussion of factors encountered in selection of pollution control systems and suggested procedures to pick the best system.

Bramer, H.C., "Pollution Control in the Steel Industry," *Environmental Science and Technology*, October 1971, Vol. 5, No. 10, pp. 1004-1008.
Presentation for the non-expert of overall pollution problems in an integrated steel-making site. Pinpoints problem areas in the overall processing scheme. Suggested methods of compliance are indicated.

Control Techniques for Sulfur Oxide Air Pollutants, U.S. Department of Health, Education and Welfare, Publication AP-52, 1969.
Sources of emission of SO_2 with methods of control including abatement equipment and modification of processes. Also presents new equipment costs.

Control Techniques for Particulate Matter, U.S. Department of Health, Education and Welfare, Publication AP-51, 1969.
Discussion of particulate matter similar to that in *Control Techniques for Sulfur Oxide Air Pollutants.*

Teller, A.J., "Control of Emissions from Glass Manufacture," *Ceramic Bulletin*, 1972, Vol. 51, No. 8, pp. 637-40.
General analysis of the particulate and flouride emissions from glass manufacture; procedures for handling the problem and rough cost estimates for a given plant size.

Additional References

The following section contains sources of data and related information used to estimate costs of pollution control equipment not cited in the text.

Reports and Books

Air Pollution Aspects of Brass and Bronze Smelting and Refining Industry, Brass and Bronze Institute, 1969.

Atmospheric Emissions from Sulfuric Acid Manufacturing Processes, United States Department of Health, Education and Welfare, Report No. 999-AP-13, 1970.

Brown, R.F., "Major Application Report # 4, A Report on Use of Electrostatic Precipitators in Rock Product Industries," *Rock Products*, May 1968, p. 120.

Compilation of Air Pollutant Emission Factors, U.S. Environmental Protection Agency, Report No. AP-42, 1972.

Drygen, C.E. and Furlow, H., *Chemical Engineering Costs*, Ohio State University, 1966.

Jownie, H.W. and Varga, J., *A System Analysis of the Integrated Iron and Steel Industry*, Batelle Memorial Institute, 1969.

Keystone Coal Industry Manual, New York: McGraw-Hill, 1969, 1970, and 1971.

Minerals Yearbook, United States Department of the Interior, Bureau of Mines, 1970.

Oglesby, S., *A Manual of Electrostatic Precipitators Technology, Part I and Part II*, Southern Research Institute, 1970.

Shreve, R.N., *Chemical Process Industries*, Third Edition, New York: McGraw-Hill, 1967.

West Virginia Coal in 1972, West Virginia Coal Association, 1972.

West Virginia Manufacturing Directory, West Virginia Department of Commerce, 1972.

Journal Articles

Akitsume, K. and Takae, T., "Abatement of Drilling Tower Effluent," Reprint, 73rd American Institute of Chemical Engineers Meeting, 1972.

Balch, G.E., "Control of Coke Oven Emissions by Use of a Closed Pipeline Charging System," *Journal of Air Pollution Control Association*, March 1972, Vol. 22, No. 3, pp. 187-90.

Besalke, R.E., "Air Pollution Control in the Refractories Industry," Reprint, 73rd AIChE Meeting, 1972.

Burch, E.F., "Coal Preparation," *Mining Congress Journal*, February 1970, Vol. 56, pp. 63-67.

Dailey, W.H., "Integrated Steel Plants of the Future," *Iron and Steel Engineer*, April 1972, Vol. 49, p. 56.

Edgar, W.D., "Coke Oven Air Emission Abatement," *Iron and Steel Engineer*, October 1972, Vol. 49, p. 86.

Hailstone, R.E., "Air Pollution Control in the Cement Industry," *Minerals Processing*, May 1969, p. 11.

Henschen, H.C., "Wet vs. Dry Cleaning in the Steel Industry," *Journal of Air Pollution Control Association*, 1968, Vol. 18, p. 328.

McIlwaine, R.W., "Air Pollution Equipment for Foundry Cupolas," *Journal of Air Pollution Control Association*, 1967, Vol. 17, No. 8, p. 540.

Northcott, E., "Dust Abatement at Bird Coal," *Mining Congress Journal*, November 1967, Vol. 53, No. 11, pp. 29-36.

Tihanksky, D.P., "A Cost Analysis of Waste Management in the Steel Industry," *Journal of Air Pollution Control Association*, 1972, Vol. 22, p. 335.

Wright, R.J., "Concepts of Electric Arc Furnace Fume Control," *Journal of Air Pollution Control Association*, 1968, Vol. 18, p. 175.

News Items

Blast Furnace and Steel Plant, February 1971, Vol. 59, p. 120; March 1971, Vol. 59, p. 168.

Business Week, February 6, 1972, pp. 52-56; July 31, 1971, p. 17.

Chemical Engineering, April 21, 1969, Vol. 76, p. 64; December 13, 1971, Vol. 78, p. 86; January 10, 1972, Vol. 79, p. 40; each issue, Marshall and Stevens Cost Index.

Chemical Week, June 30, 1971, p. 15; August 25, 1971, p. 14; September 8, 1971, p. 31; September 8, 1971, p. 40; October 13, 1971, p. 22; October 27, 1971, p. 33; November 10, 1971, p. 65; November 17, 1971, p. 24; November 17, 1971, p. 65; December 8, 1971, p. 34; March 15, 1972, p. 11; March 29, 1972, p. 16; May 17, 1972, p. 20; August 16, 1972, p. 21; September 13, 1972, p. 14; August 22, 1973, p. 16.

Environmental Health Letter, May 1, 1972, p. 2.

Environmental Science and Technology, August 1971, Vol. 5, pp. 662-65; September 1971, Vol. 5, p. 749; October 1971, Vol. 5, p. 989; February 1972, Vol. 6, p. 107; February 1972, Vol. 6, p. 118; July 1972, Vol. 6, p. 585; November 1972, Vol. 6, pp. 971-73.

Iron Age, August 24, 1972, Vol. 210, p. 27.

Journal of Air Pollution Control Association, 1969, Vol. 19, p. 278; 1973, Vol. 23, No. 5, p. 406.

Morgantown Dominion-News, March 1, 1973, p. 8.
The New York Times, April 22, 1971, p. 57, section 3; May 14, 1971, p. 18, section 1; August 3, 1972, p. 41, section 2.
Wall Street Journal, March 13, 1972, p. 3.

Index

Index

About the Authors

William H. Miernyk received the B.A. from the University of Colorado in 1946 and the M.A. and the Ph.D. from Harvard University in 1953. He is Benedum Professor of Economics and Director of the Regional Research Institute at West Virginia University, Morgantown. He has taught at Northeastern University and the University of Colorado, and was visiting Professor of Economics at the Massachusetts Institute of Technology in 1957/58 and at Harvard University in 1969/70. Dr. Miernyk has served as consultant to the U.S. Senate Committee on Commerce, and to advisory committees in the U.S. Departments of Commerce and Labor. He is coauthor of *The Economic State of New England* (Yale University Press, 1954) and *Simulating Regional Economic Development* (Lexington Books, D.C. Heath and Company, 1970); he is author of *Inter-Industry Labor Mobility* (Northeastern University, Bureau of Business and Economic Research, 1955); *Trade Unions in the Age of Affluence* (Random House, 1962); *The Economics of Labor and Collective Bargaining* (D.C. Heath, 1965); *The Elements of Input-Output Analysis* (Random House, 1965); and *Economics* (Random House, 1971). Dr. Miernyk has contributed to twelve additional books, and is author of more than one hundred monographs, articles, and reviews.

John T. Sears received the B.S. from the University of Wisconsin in 1960 and the Ph.D. from Princeton University in 1965. He is Associate Professor of Chemical Engineering at West Virginia University, where he has taught since 1969. Prior to that, Dr. Sears was a research chemical engineer at the Brookhaven National Laboratory and at the Esso Research and Engineering Center. He is author of eighteen articles and technical papers. Dr. Sears received the Dow Outstanding Young Faculty Award in 1971.